Palgrave Studies in Environmental Sociology and Policy

Series editor
Ortwin Renn
Director at the Institute for
Advanced Sustainability Studies (IASS)
Potsdam, Germany

This series is dedicated to environmental sociology as a distinct field with emphasis on theoretical grounding, empirical validation, evidence–based research and practical policy application. The series will include studies that focus on state-of-the-art research and an explicit effort to transfer the results into recommendations for policy makers in their quest for aknowledge-driven transformation towards sustainability. The series will incorporate combinations of conceptual analyses with clear applications to environmental policy making and empirical studies that have the potential to inform decision and policy makers from government, public agencies, private business and civil society.

More information about this series at
http://www.palgrave.com/series/15613

Annika Arnold

Climate Change and Storytelling

Narratives and Cultural Meaning in Environmental Communication

Annika Arnold
Universität Stuttgart
Stuttgart, Germany

Palgrave Studies in Environmental Sociology and Policy
ISBN 978-3-319-69382-8   ISBN 978-3-319-69383-5 (eBook)
https://doi.org/10.1007/978-3-319-69383-5

Library of Congress Control Number: 2017959560

© The Editor(s) (if applicable) and The Author(s) 2018
This work is subject to copyright. All rights are solely and exclusively licensed by the Publisher, whether the whole or part of the material is concerned, specifically the rights of translation, reprinting, reuse of illustrations, recitation, broadcasting, reproduction on microfilms or in any other physical way, and transmission or information storage and retrieval, electronic adaptation, computer software, or by similar or dissimilar methodology now known or hereafter developed.
The use of general descriptive names, registered names, trademarks, service marks, etc. in this publication does not imply, even in the absence of a specific statement, that such names are exempt from the relevant protective laws and regulations and therefore free for general use.
The publisher, the authors and the editors are safe to assume that the advice and information in this book are believed to be true and accurate at the date of publication. Neither the publisher nor the authors or the editors give a warranty, express or implied, with respect to the material contained herein or for any errors or omissions that may have been made. The publisher remains neutral with regard to jurisdictional claims in published maps and institutional affiliations.

Cover illustration: © saulgranda/Getty

Printed on acid-free paper

This Palgrave Macmillan imprint is published by Springer Nature
The registered company is Springer International Publishing AG
The registered company address is: Gewerbestrasse 11, 6330 Cham, Switzerland

Acknowledgments

This work was made possible by a scholarship provided through the Friedrich-Ebert-Foundation Germany. The book would not have happened without the cultural and institutional support offered by the University of Stuttgart and the Center for Cultural Sociology, Yale University. Institutions are nothing without the people that fill their halls with life and intellect, so my special thanks goes to my supervisors, Ortwin Renn, Stephan Moebius and to Philip Smith as well as to colleagues and friends along the way: Shai Dromi, Jin Su Joo, Stina Kjellgren, Hannah Kosow, Marco Sonnberger and Jette Weiss. On the publishing side, I want to thank Rachael Ballard and the team at Palgrave Macmillan for their advice, encouragement and patience. Empirical research is nothing without those who provide knowledge, insights and not least their time to the researcher, so I want to thank the interviewees who contributed to this project.

I also want to thank Veronika Arnold and Gerd Prandtstetter for their support and a special thank you to Karin and Ulli Arnold, my parents, for their constant encouragement in every way imaginable. Their contribution to this project is more than I can comprehend.

Contents

1 Introduction: Why Narratives Matter in Climate Change Communication   1

2 Climate Change Communication Studies: Inquiries into Beliefs, Information and Stories   7

3 How to Understand the Role of Narratives in Environmental Communication: Cultural Narrative Analysis   57

4 Telling the Stories of Climate Change: Structure and Content   83

5 Conclusions: Pitfalls and the Power of Narratives   123

Index   135

LIST OF FIGURES

Fig. 2.1 Opposing arguments in the climate change discourse 14
Fig. 2.2 Down's issue attention cycle 16
Fig. 2.3 Social ecology – the interaction between cultural and natural spheres of causation 24
Fig. 2.4 Grid-group dimensions and the five myths of nature 30
Fig. 3.1 The structural model of genre 71
Fig. 3.2 Integrated model of cultural narrative analysis (own illustration) 77
Fig. 4.1 Climate advocates' narratives (own illustration) 86

LIST OF TABLES

Table 2.1	Paradigms, problems, and proposals [in 25 years of PUS research]	21
Table 2.2	Overview: terminology and topics in climate change communication literature	35
Table 3.1	The social role of narratives	61
Table 3.2	Narrative in linguistics, literary theory, and social sciences	64

CHAPTER 1

Introduction: Why Narratives Matter in Climate Change Communication

Abstract To understand the nature of the political and public debate about climate change, we need to understand the narrative structures that produce this discourse. Narratives, occurring in media, in public discourse, political agenda or even scientific debate, are vehicles for complex phenomena, such as climate change. The science behind this, the intricate interrelations and differences between daily weather occurrences and climate, between various factors – from natural causes to a changing climate to anthropogenic climate change – and the sheer amount of voices in this debate, make climate change a hard topic to sell. Climate change policies are complicated and in need to factor in a large amount of different aspects. As "story-telling animals", we perceive facts, numbers and urgent appeals that surround climate change inherently as a story.

Keywords Climate change communication • Storytelling • Narratives • USA • Germany • Communication research

We grow up listening to and telling stories. In fact, every night, as millions of moms and dads put their kids to bed, they will read books to or tell tales to those kids. And in turn, those moms and dads, just like everyone else, reflect their lives and the events that are happening, big or small, in the structure of a story. Narrative structures allow us to gain understanding of

© The Author(s) 2018
A. Arnold, *Climate Change and Storytelling*,
Palgrave Studies in Environmental Sociology and Policy,
https://doi.org/10.1007/978-3-319-69383-5_1

events and how they relate to one another and to our lives. This is not something we learn at school, but something we already experience as small children (Alda 2017: 164–165). As Alexander puts it: we are all story-telling animals (Alexander 2003: 84). A good story can be suspenseful, engaging, building a sense of belonging, give meaning and order to a complex world. And it can be persuasive. This book is about such stories, trying to make sense of one of the biggest environmental crises we have faced so far: climate change. And it is about those who tell these stories to fight climate change: environmental activists, politicians, civil society actors, and even artists. Climate change is a highly complicated problem; it is made up of complex climatological data, of economic reasoning, empathy for others and day-to-day and long-term politics. It touches on the kind of society we live in, its logic and its shortfalls. But in order to make the fight against climate change a priority, climate advocates need to tell stories, to mobilize people and guide their actions (Shenhav 2015: 5). In its 2016 flagship report, the German advisory council on global change called for narratives of and for change (WBGU 2016) to encourage innovations and to connect with the cultural fabric of society. But the current discourse about sustainability and changing environmental conditions is not taking place in a vacuum. It is already spun into a web of meaning, the problem gets translated into a story with all the required elements: heroes, villains, victims, an object of struggle, a beginning, middle, end, and morale of the narrative.

Climate change as a macro-environmental issue meets the criteria of "super-wicked problems" (Lazarus 2009: 1159), i.e. problems that are characterized by uncertainty over consequences, diverse and multiple engaged interests, conflicting knowledge, and high stakes. Climate, as a common pool resource (Renn 2011), poses one of the most pressing policy problems our society is facing today. However, climate change seems to be the first environmental crisis in which experts appear more alarmed than the public. "People think about 'global warming' in the same way they think about 'violence on television' or 'growing trade deficits', as a marginal concern to them, if a concern at all" (Hamblyn 2009: 234). The impacts of a changing climate are hard to grasp and solutions to the problems are diverse, complex, and controversial. Public perception of associated risks plays a huge role when it comes to support for climate policies and this perception is culturally determined: "Culture affects how humans understand the world, because we make sense of the world by cultural means" (Arnoldi 2013: 107). Berger and Luckmann ([1990], c1966) famously stated how our reality is the result of social construction, a collective effort to make sense of the world as we see it. The way we construct this reality by means of social communication has been subject to a wide range of sociological research. Goffman (2010) introduced framing as a means to read and understand situations and activities in social life. Helgeson et al. stress the role of cognitive structures in his concept of a mental model, which is "a person's internal, personalized, intuitive, and contextual understanding of how something works" (Helgeson et al. 2012: 331). Boholm provides the concept of culture as shared schemata that allow us to process meanings and order information due to defined categories, relationships and contexts (Boholm 2003: 168).

This book analyzes narratives in qualitative data – interviews conducted with US-American and German climate advocates, i.e. people who are dedicated to fighting climate change and to engaging and motivating people around them. The analysis draws on existing narrative theory and suggestions for narrative analysis. The aim is to add to the understanding of environmental communication, especially in the field of climate change. Much existing discourse analysis addressing the topic of climate change focuses on media representation of the discourse (Boykoff 2008; Boykoff and Boykoff 2004; Downs 1972), the debate between climate sceptics and climate advocates (Hoffman 2011; The Pew Research Center 2007), or takes an instrumental stand on the issue by asking how climate change communication should look like in order to achieve agreements (manipulate) within the civil society (Hart and Nisbet 2011; Moser and Dilling 2011; Moser 2010). Focusing on media representation is valuable to see which information the wider society gains, however, this field analyzes communication elements that have already been processed and are shaped according to the rules of the media landscape. Focusing on the tensions and arguments made between climate advocates and their opponents enhances factual understanding of pro-/ con arguments and might help to address them properly (if one's goal is to better climate change communication). But this approach neglects that mistakes have been made in the communication process before the pro-climate arguments are re-told by the media. A purely instrumental in-order-to-approach won't reveal the cultural process of civil discourse, because it is too strongly focused on providing recipes for communication handbooks. With the help from cultural theory social sciences can contribute to the understanding of environmental communication by considering that communication processes are not at all specific to one subject but follow inherent rules that need to be uncovered. For this, narrative analysis can "help investigators think about 'non-rational' characteristics of environmentally relevant situations" (Shanahan et al. 1999: 417).

Using This Book

This book is on the one hand a coherent story itself, making the case for strong narrative research in the social sciences and presenting empirical findings to exemplify how such research efforts could look like. On the other hand, the different chapters can be used on their own and serve different purposes: the second chapter, that deals with social sciences' research into the topic of climate change proposes a starting point for those who are interested in the social and cultural sciences' role in climate communication and environmental studies. The third chapter on narrative analysis might be of interest for students who are just starting out in the field of narrative analysis. It provides an in-depth overview of the origins of narrative analysis and application examples. The fourth and fifth chapter presents and discusses findings in the empirical data conducted for this study only. It suggests a blueprint on how to make use of narrative theory for understanding stories and their use for environmental communication.

This book is not a cookbook for communication strategies; you will not find direct recommendations, claiming to put this topic or that topic at the center and one actor group as a villain and another as a hero. What you will hopefully take away from this book are insights into the art and power of telling stories about real-life events and the pitfalls that come along with them. Mostly, instead of focusing on the audience and trying to figure out what a specific audience might want to hear, this study goes where those stories originate, before they are observed by an audience: the people who are telling these stories, their reasoning and their struggles. This book tells the story of the storyteller.

References

Alda, A. (2017). *If I Understood You, Would I Have This Look on My Face? My Adventures in the Art and Science of Relating and Communicating*. New York: Random House.

Alexander, J. C. (2003). On the Social Construction of Moral Universals: The "Holocaust" from War Crime to Trauma Drama. In J. C. Alexander (Ed.), *The Meanings of Social Life. A Cultural Sociology* (pp. 27–84). Oxford/New York: Oxford University Press.

Arnoldi, J. (2013). *Risk* (1st ed.). New York: John Wiley & Sons.
Berger, P. L., & Luckmann, T. ([1990], c1966). *The Social Construction of Reality. A Treatise in the Sociology of Knowledge.* New York: Anchor Books. Available Online at http://www.loc.gov/catdir/description/random0414/89018142.html
Boholm, Å. (2003). The Cultural Nature of Risk. Can There Be an Anthropology of Uncertainty? *Ethnos, 68*(2), 159–178. https://doi.org/10.1080/0014184 032000097722.
Boykoff, M. T. (2008). Lost in Translation? The United States Television News Coverage of Anthropogenic Climate Change, 1995–2004. *Climatic Change, 86,* 1–11.
Boykoff, M. T., & Boykoff, J. M. (2004). Balance as Bias: Global Warming in the US Prestige Press. *Global Environmental Change, 14,* 125–136.
Downs, A. (1972). Up and Down with Ecology – The "Issue-Attention-Cycle". *The Public Interest, 28,* 38–51.
Goffman, E. (2010). *Frame Analysis. An Essay on the Organization of Experience.* Boston: Northeastern University Press.
Hamblyn, R. (2009). The Whistleblower and the Canary: Rhetorical Construction of Climate Change. *Journal of Historical Geography, 35,* 223–236.
Hart, P., & Nisbet, E. C. (2011). Boomerang Effects in Science Communication: How Motivated Reasoning and Identity Cues Amplify Opinion Polarization About Climate Mitigation Policies. *Communication Research.* https://doi.org/10.1177/0093650211416646.
Helgeson, J., van der Linden, S., & Chabay, I. (2012). The Role of Knowledge, Learning and Mental Models in Public Perceptions of Climate Change Related Risks. In A. E. J. Wals & P. B. Corcoran (Eds.), *Learning for Sustainability in Times of Accelerating Change* (pp. 329–346). Wageningen: Wageningen Academic Publishers.
Hoffman, A. J. (2011). The Culture and Discourse of Climate Skepticism. *Strategic Organization, 9*(1), 77–84.
Lazarus, R. J. (2009). Super Wicked Problems and Climate Change: Restraining the Present to Liberate the Future. *Cornell Law Review, 94*(5), 1153–1234.
Moser, S. C. (2010). Communicating Climate Change: History, Challenges, Process and Future Directions. *Wiley Interdisciplinary Reviews: Climate Change, 1*(1), 31–53. https://doi.org/10.1002/wcc.11.
Moser, S. C., & Dilling, L. (2011). Communicating Climate Change: Closing the Science-Action Gap. In J. S. Dryzek, R. B. Norgaard, & D. Schlosberg (Eds.), *Oxford Handbook of Climate Change and Society* (pp. 161–174). Oxford/New York: Oxford University Press.
Renn, O. (2011). The Social Amplification/Attenuation of Risk Framework: Application to Climate Change. *Wiley Interdisciplinary Reviews: Climate Change, 2*(2), 154–169.

Shanahan, J., McComas, K., & Pelstring, L. (1999). Using Narratives to Think About Environmental Attitude and Behavior: An Exploratory Study. *Society & Natural Resources, 12*(5), 405–419. https://doi.org/10.1080/089419299279506.

Shenhav, S. R. (2015). *Analyzing Social Narratives. 1. Auflage*, Routledge Series on Interpretive Methods. New York [u.a.]: Routledge.

The Pew Research Center. (2007). *Global Warming: A Divide on Causes and Solutions*. Public Views Unchanged by Unusual Weather. Washington, DC.

WBGU - German Advisory Council on Global Change. (2016). *Der Umzug der Menschheit: die transformative Kraft der Städte*. Berlin: WBGU.

CHAPTER 2

Climate Change Communication Studies: Inquiries into Beliefs, Information and Stories

Abstract This chapter provides an overview of selected aspects of social sciences' studies on the phenomenon of climate change. In particular, this chapter discusses studies in the realm of risk perception and risk communication; it pays specific attention to insight from discourse analysis and findings in media research, such as the norm of balanced reporting. Scholarship of public understanding of science and science communication provides additional information on how societies perceive the risk of climate change. With this in mind, the chapter closes with a closer look at cultural theory and cultural study approaches to the topic before introducing the cultural sociological perspective as the theoretical basis for the following empirical analysis.

Keywords Climate change communication • Media analysis • Discourse analysis • Risk communication • Cultural sociology • Cultural theory • Public understanding of science

Before diving into an application of narrative theory for a cultural-sociological take on climate change communication, it is worth to lay out previous research in neighboring fields that have significant influence here: from media studies to risk communication, from inquiries into public understanding of science, discourse analyses and cultural studies approaches –

communication itself is a multi-layered phenomenon and in combination with a "super-wicked-problem" (Wiesenthal 2010: 184–185) like climate change, its analysis should make use of all those different approaches. The body of significant literature has grown enormously over the last years; to review and discuss every single strand in this field would go beyond the scope of this study, so the following chapter will rather highlight some of those studies that are directly connected to the analysis in later chapters.

Risk Perception and Risk Communication

Environmentalists and commentators across the board agree – to a large part – at least on one specific challenge when it comes to addressing climate change: it feels far away, both in terms of time and in terms of space. This is a specific interesting conundrum for risk perception scholarship, addressing questions like:

- How do people perceive and estimate the risks related to global warming (Whitmarsh 2008; Leiserowitz 2005, 2007)?
- How do people react to the uncertainties within climate change research (Renn 2008)?

If people are faced with uncertain consequences of risks and if they do not have the resources to address these risks properly, they tend to resolve this cognitive dissonance in order to go on with their every-day life (Aronson 2008). This observation has to be considered when talking about successful ways of climate change communication examining the role of fearful messages. Operating with fear as a motivational tool can be risky: it is difficult to sustain fear in the long term because the audience might become desensitized to fear appeals. Drawing on results from social psychology and behavior studies, scholars point out dangers that lie in using fear as a motivator. Painting overly dramatic pictures of doom and the devastating effects of climate change might draw public's attention for a short time, but if clear and applicable guidelines are not provided people are only left helpless and scared (Moser and Walser 2008) and – confronted with a sheer irresolvable challenge – will retreat and disengage completely (Ereaut and Segnit 2006; Hamblyn 2009: 235). Fear appeals are likely to jeopardize audience's trust in those organizations that use fear in their messages and fear appeals might lead to unintended reactions (O'Neill and Nicholson-Cole 2009: 360–361). Aronson points out another danger fear messages carry exemplifying this by an experiment conducted among students at the University of California, Los Angeles in 1986. College freshmen have been informed about the likelihood of an earthquake taking place in the Los Angeles area and then were assigned randomly to different dormitories. One half moved into newly build houses that were relatively safe seismically, the other half lived in older, more vulnerable buildings. After some weeks researchers asked both groups about their knowledge on earthquakes and surprisingly found that those students living in the older dorms had significantly less knowledge on appropriate behavior in case of an earthquake than the other 50% of students. What happened? Aronson argues that even though the students were plenty scared when they first heard about the likelihood of an earthquake they were not provided with strategies. So in order to go on with daily life, students started to play down the chance of becoming victim to natural disasters. They thus dissolved the cognitive dissonance they faced; simply going into denial about the problem. This experiment shows that knowledge and information alone will not lead to a behavioral change; on the contrary, it might even become counterproductive if no guidance is provided. Recommendations for taking action have to be effective, concrete, and doable, otherwise a message of fear will not produce reasonable responses to danger, but instead it will produce denial (see also cf. Moser and Dilling 2011: 40 and Moser and Dilling 2004). In this scenario, the threat of an earthquake is only a vague idea; most students may never have been victim to natural disasters. Maybe the experience has to be more drastically life-like than pure information can be? Some scholarship (e.g. O'Neill and Nicholson-Cole 2009; Moser and Dilling 2011) object the use of fear loaded messages since audiences often reject fear messages as manipulating and attention-grabbing messages do not necessarily empower action. Similar to Aronson, O'Neill argues that fear messages, even though they capture a lot of attention via shocking and sensational pictures, may leave people feeling helpless and overwhelmed, so going into denial or frustration with the issue is a likely outcome (O'Neill and Nicholson-Cole 2009: 374). If fear messages and dangerous threats do not have the "right" outcome, then what about the actual experience of natural disasters? Whitmarsh (2008) investigates this question by intersecting the question of the role of knowledge and experience to the question of risk perception, examining people's perception of climate change after they were victims to flood events, which may or may not be caused by the impacts of global warming. The author examines the role of direct experience in perception of and individual response to the likely outcomes of global warming. The data for this study was conducted in the UK in 2003 and was based on the hypothesis that people who already experienced flooding and damages to their possessions will pay more attention to the possibilities and dangers of future floods. This thesis is based on the assumption that direct and personal experiences influence individual risk perception (Bickerstaff and Walker 1999), that attention towards a risk rises if this risk has already been personally experienced (Keller et al. 2006), and that people perceive local risks as more threatening (Hinchliffe 1996). However, the author concludes no significant differences of knowledge about global warming between flood-victims and non-flood-victims. There was also no significant difference concerning response and perception of possible impacts of climate change. Whitmarsh offers an explanation for the rejection of the hypothesis that floods and climate change are perceived as somewhat divergent issues: floods are directly experienced by those who were damaged and require a sudden and urgent solution; climate change however is a long-term problem with solutions that are less simple and obvious. The time-spatio distance of climate change's consequences makes it thus especially difficult in every-day life to worry enough about the issue to actually alter human behavior. The high level of uncertainty that surrounds climate change leads to an indefinite dimension of action, preventing a significant and sustainable change in cultural lifestyle. Even though respondents in surveys often attribute great importance to the topic of environmental concerns, acting accordingly does not follow. This phenomenon is characterized by the value action gap (Owens and Driffill 2008; Kollmuss and Agyeman 2002), or, as Giddens calls it: the Giddens' Paradox (Giddens 2009: 2; 113). This value-action-gap poses a dilemma to psychological as well as to sociological scholarship and carries important clues for climate change communication. Here, the question is why people do not act according to their – survey-tested – knowledge (Leggewie and Welzer 2010: 74). One explanation lies within the spatio-temporal divergence between causes and impacts. Kuckartz identifies three dimensions that influence action – or rather – non-action (Kuckartz 2010: 151):

- The dimension of space: people will only alter their behavior if the consequences of that behavior are spatially directly tangible. Kuckartz argues that since there are no real severe impacts of climate change to be feared in Germany, willingness to pay is comparably low.
- The dimension of time: people tend to prioritize needs alongside the degree of urgency among other factors. As Giddens puts it: "People

find it hard to give the same level of reality to the future as they do to the present" (Giddens 2009: 2).
- The dimension of community and the tragedy of the commons (Ostrom 1990; see also Renn 2011: 161–165): climate mitigation is non-exclusive to those who put in the effort. People also justify their lack of taking action with reference to the limitation of their scope, pointing out that industry and politics have to take on the vanguard role in order to achieve actual results.

In his widely noticed "The Politics of Climate Change" Anthony Giddens applies the basic idea of the value-action-gap on macro-level reflections, concerning himself solely with the history and current state of climate, environmental, and energy politics.

> The Giddens' paradox lies in the observation that "no matter how much we are told about the threats, it is hard to face up to them, because they feel somehow unreal – and, in the meantime, there is a life to be lived, with all its pleasures and pressures. The politics of climate change has to cope with what I call 'Giddens' paradox'. It states that, since the dangers posed by global warming aren't tangible, immediate or visible in the course of day-to-day life, however awesome they appear, many will sit on their hands and do nothing of a concrete nature about them. Yet waiting until they become visible and acute before being stirred to serious action will, by definition, be too late. (Giddens 2009: 2)

A specific sociological observation lies beneath these explanatory factors: the observation that attitudes and norms are not just the primary guiding principles but context of action has to be considered as well (Leggewie and Welzer 2010: 74–79). The question why people do not act according to their knowledge thus has to be answered with regard to individual as well as social barriers (Nicholson-Cole 2005; O'Neill and Hulme 2009). Hence, we are able to handle our value system the way the situation requires it, for example a justification for a decision can be brought into accordance with our values and norms by considering extraordinary circumstances.

Experts who want to alert an audience to a topic like climate change have to take into account all these different levels of engagement and interests and, like it is with almost every complex, scientific topic, they have to deal with different level of knowledge about the issue.

Discourse-Analytic Studies: Media Logic, Scientific Reporting and Science Communication

Mass media provide a way to introduce scientific facts and knowledge into the broader social discourse and to gain attention for the topic of concern. The analysis of societal discourse has become ever more interesting to climate change scholarship. Studies in this field primarily focus on the discourse taking place in public media:

> In research on the public understanding of climate change, we operate under the global hypothesis that cycles in media coverage embody narratives that guide public understanding. [...] Communication research has delved into the narratives told by the mass media, which some argue have emerged as modern society's primary storytellers, having in many ways replaced society's dependencies on direct information and oral traditions. (Trumbo and Shanahan 2000: 201)

In her 2015 book "Risk and anthropology", Åsa Boholm sees collective narratives about events as communicated through news media (Boholm 2015).[1] News coverage of climate change – or any other topic for that matter – plays an important role in bringing the topic into the civil sphere, elucidating the issue to laypeople, and producing political pressure. Here, key questions are:

- To what extent is climate change covered by the media (e.g. Schmidt et al. 2013; Schäfer et al. 2011; Weingart et al. 2000)?
- What does media coverage of climate change look like (e.g. Dahl 2015; Pearce et al. 2015; Myers et al. 2012; Maibach et al. 2010; Nisbet 2009; Ereaut and Segnit 2006)
- Which challenges does climate change pose for the realm of science communication (e.g. Hart and Nisbet 2011; Bauer et al. 2007; Downing and Ballantyne 2007; Nisbet and Goidel 2007)?

Trumbo identifies climate change as an outstanding example of new environmental problems. Thus, climate change poses a challenge to media coverage which has to deal with its specifics of being a long-term issue with global consequences, while at the same time its 'happening' stays invisible to most of the audience (Trumbo 1996: 269). Knowledge about climate change is mostly distributed and perceived through mass media (Nelkin 1987 and Wilson 1995, all cited in: Lowe et al. 2006: 436).

Looking into news coverage regarding climate change it becomes obvious that the logic of media reporting and the logic of scientific reporting do not always coincide. There is a large body of work on media perception, reporting, gatekeeper influence, and communicator-recipient-relation (Gans 2005; Bell 1991; Tuchman 1978; Galtung and Ruge Mari 1965). Here, I will particularly report the results of research that examines how the issue of global warming is treated in the media. A first step is to examine how much attention media are paying to the topic, i.e. how much space environmental topics get in newspapers and how often reports occur (see for example: Weingart et al. 2000). Schäfer et al. (2011) report 6.894 articles appeared in German newspapers between 1996 and 2010.[2] Interestingly enough, the focus in media studies seems to lie within print media, while online news or even TV shows are less frequently evaluated. An exception is Boykoff and Boykoff's (2007) study of U.S. American television news coverage on global warming. Some analyses even compare numbers of articles, wording, and journalistic practices in an international context (Boykoff 2008; Nissani 1999; Mazur 1998; Bell 1994).

Media reporting follows different rules than scientific reporting, thus it is likely that the case of climate change gets presented in a different way as it occurs in the professional debate. Messages thus need to be tailored to a specific medium and its designated audience, "using carefully researched metaphors, allusions, and examples that trigger a new way of thinking about the personal relevance of climate change" (Nisbet 2009: 15). In public debates, issues get organized according to the frame they best fit, which, in turn, depends on the 'debate genre'. Within politics and lobbyism issues are framed in order to reach decisions and to identify policy options. On a technical expert level, frames need to present technical details as more accessible to enable experts on a social level to turn them into persuasive stories. And journalists use frames to present interesting news reports (Nisbet 2009). Journalistic work is not merely the distribution of scientific knowledge. It rather underlies various cultural influences, political expectations, and narrative requirements. Variations in media's reinterpretation of climate change facts can be explained by considering the role of value orientation. Carvalho and Burgess (2005) follow up this assumption with a content analysis of climate change reports in three UK newspapers, covering the release of the IPCC's assessment report in 1995. The authors describe how "The Independent" and "The Guardian" followed the conclusions of the IPCC[3] report and mobilized public concern, while The Times attempted to discredit the work of the IPCC and to persuade its readers that climate change did not present any significant risk to society. This study shows "that coverage of climate change has been strongly linked to the political agenda on this issue, and particularly to public pronouncements and discursive strategies of prime ministers and other top governmental figures" (Carvalho and Burgess 2005: 1467). Aside from political interests that can be either promoted or hindered, reports on climate change are often accompanied by inaccurate scientific facts and a sense of alarmism (Carvalho and Burgess 2005; Post 2008). This sense of alarmism is underlined in a report by the institute for public policy research (Ereaut and Segnit 2006), where 10 different linguistic climate change repertoires within newspapers are found. The sheer number of repertoires alone shows the key finding of a contradictory, confusing, and chaotic discourse (Fig. 2.1).

The debate in UK mainstream media is filled with different voices and contradicting opinions, with a lean towards the left side, where climate change is perceived as anthropogenic, massive, and real. However, the authors summarize different repertoires as three categories, one with an overall pessimistic perspective, containing an Alarmist repertoire, and two others: the optimistic repertoires and the pragmatic optimistic repertoires. The latter are ascribed as "it will be alright" attitude, whereas the first characterized by a doom-perspective, perceiving climate change as the

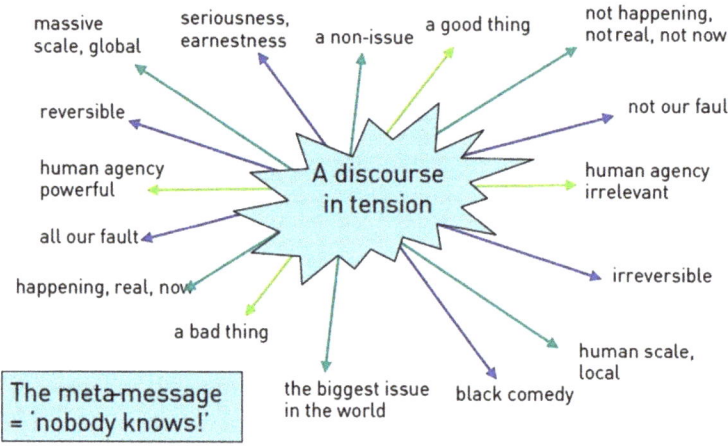

Fig. 2.1 Opposing arguments in the climate change discourse (Source: Ereaut and Segnit 2006: 10)

biggest challenge the world is facing today and seeing the earth passed the point of no return (Ereaut and Segnit 2006: 13).

Storch and Krauss undertake a transcultural analysis of media coverage in U.S. American and German newspapers and argue that "even though there are significant differences in the public understanding of climate change [...], the media in both societies use a similar framework of vulnerability, even if it is constructed in culturally different ways" (Krauss and Storch 2005: 2). These cultural differences are especially found in varying wording and the implications of that choice of wording, the U.S. American media referring with the term 'global warming' to a tendency toward warmer mean temperatures while German media's 'Klimakatastrophe' pictures climate change as a somewhat broader term, emphasizing a construction between disastrous weather events and the consequences of a changing climate. The authors exemplify this on the report of the Elbe River flooding in August 2002, where a regional newspaper reported that "now the flood finally reached our backyards. This flood confronts us with the 'why', with the sins we have committed, with the search for its origins. Even without scientific certainty we know that the flood is a consequence not only of cosmic changes, but of our own way of living" (Krauss and Storch 2005: 3). Schäfer et al. (2011) and Grundmann (2007) identify specific events as points of time that lead to an increase of reports on global warming, such as international climate summits like the Copenhagen Summit in 2009 or when a new IPCC report is released. Those events draw public attention towards environmental problems and for a short period of time, climate negotiations about CO_2 reductions and cap and trade are 'breaking news'.

Media Coverage of Climate Change I: The Issue Attention Cycle

According to Downs' issue attention cycle news coverage on these events leads to a different relation to public attention. It is what he calls the post-problem stage, the 5th phase within the cycle, which mostly refers to public attention towards ecological problems. Downs identifies three characteristics inherent to those problems that eventually pass through the entire issue attention cycle; all of which are a perfect fit for environmental issues: only a numerical minority is suffering from the problem (A); causes for those sufferings are largely due to benefits for a majority (B); the problem is not able to deliver dramatic and exciting footage, like pictures and videos, to compete with other news (C) (Downs 1972: 41) (Fig. 2.2).

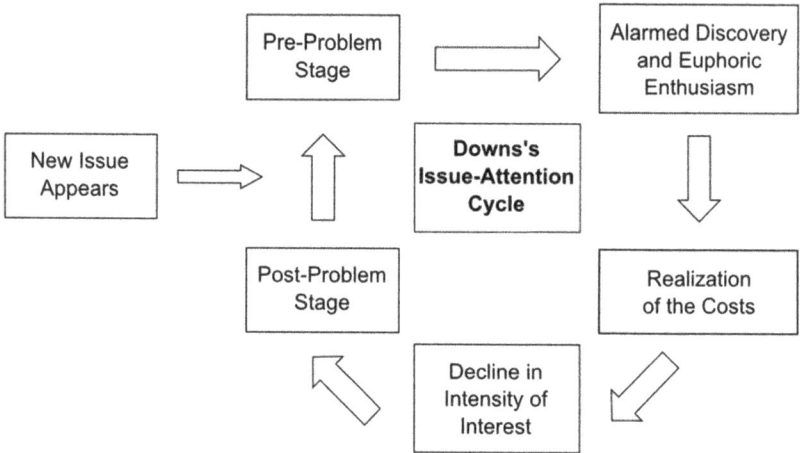

Fig. 2.2 Down's issue attention cycle (Source: Petersen 2009: 7)

The cycle starts with the pre-problem stage (1), where an undesirable social condition exists, but only alarms experts or specific interest groups, while public attention is not yet focused on this issue. In stage two, 'Alarmed discovery and euphoric enthusiasm' (2), public attention rises highly due to a series of events that bring about the specifics and harmful consequences of a problem. Calls for immediate and effective action are being made, often on the political level. Those claims are getting fewer when stage three hits: 'Realizing the cost of significant progress' (3). Public opinion, often via media reports, starts to understand the financial and maybe social sacrifices that have to be made to make change happen, which leads to a decline of intense public interest (4), due to three reactions: having the solution's high cost in mind some people get discouraged, others try to avoid the problem because they feel threatened but helpless in the face of the challenge. Still others simply become fed up with the issue. What follows is aforementioned stage 5, where other issues of national or international interest capture the public's attention. However, institutional structures which were created in the wake of debate about the problem are not simply vanishing again. Thus, social debate does not just return to pre-problem status quo, but keeps the once heatedly debated issue in the loop, although on a lower level (Downs 1972: 40–41).

The model provides valuable insights into societal mechanisms which are involved in news processing. The identification of five stages can be used in media studies to enrich the analyses with a systemizing tool that might present occurrence of climate change reports in a different light and helps to portray a topic's media career (Trumbo 1996: 274). Trumbo takes the issue attention cycle as basis for an examination of newspaper coverage on global warming in five major national newspapers, The New York Times, The Washington Post, The Los Angeles Times, The Christian Science Monitor, and The Wall Street Journal. He identifies two pivotal points in time that influenced the media life of the issue: "mid-1988 when Hansen testified before Congress and mid-1992 when the Earth Summit concludes" (Trumbo 1996: 272). The author traces the five stages of Downs' model in news coverage of climate change. Although different intervals agree with the model's predictions, like the pre-problem stage that lasted until Hansen's Congressional testimony,[4] ringing in the second stage of alarmed discovery when media coverage escalated and politicians attended to the matter, the author concludes that the predicted higher-level attention at the end of the circle did not occur. "The amount of media attention during the first six months of 1993 is similar in volume to that of the first half in 1988" (Trumbo 1996: 275).

However, the model also has been criticized for being too restrictive, linear, or omitting additional factors, oftentimes from the realm of media analyses (Grundmann and Krishnamurthy 2010; Brossard et al. 2004; McComas and Shanahan 1999). McComas and Shanahan (1999) examine in a quantitative content analysis the amount of articles dealing with global warming in The Washington Post and The New York Times between 1980 and 1995, and reject Downs' model as too restrictive by pointing out that social phenomena do not move linearly from stage to stage but are subject to a dynamic change of external social factors.[5] Concerning the characteristics Downs identifies, the authors claim those features to be not specifically inherent to features inherent to environmental problems. Thus, these characteristics do not make them subject to the issue attention cycle, but "these features were also narratively constructed by media covering the issues" (McComas and Shanahan 1999: 9). Brossard et al. (2004) support this critique in their cross-cultural comparison of newspaper coverage of global warming in France and the USA, covering editions between 1987 und 1997. The characteristics of the media attention cycle could not be found in French news covering, but seems to only apply to U.S. American media. The way of telling the story about climate change in the news, however, seems to a bigger extent influenced by journalistic culture.[6] French journalists used a different set of protagonists than U.S. American media: French newspapers emphasized the conflicts between the USA and Europe, whereas U.S. American media focused on the debate between scientists and politicians, with a clear tendency towards domestic politics. This finding gives way to the principle of balanced reporting leading to a biased portrayal of the ongoing debate about climate change. Taking off from the results from Brossard's study, Grundman and Krishnamurthy (2010) amplify the study's scope by adding the public discourse of two European countries, Germany and the UK. They state that they could not find any evidence for Downs' issue attention cycle, but detect an overall rise in news coverage of Germany, the USA, UK, and France over two decades and an excessive rise after 2004.

Media Coverage of Climate Change II: The Norm of Balanced Reporting

With new data from two more countries the authors confirm Brossard's et al. (2004) diagnosis of culturally different reporting on the issue, supporting the observations made for France and assigning similar results to German and British newspapers. European media debate is characterized by an emphasis on political implications of environmental problems, with frequent references to the U.S. American role in the process. U.S. American media focuses more on the ongoing scientific debate about existence, nature, and cause of global warming. Considering the large consensus among climatologists it seems curious that a huge proportion of media reports would still concern itself with the search for evidence for an anthropogenic climate change.

However, it does fit into the logic of media discourse. Especially in the USA, journalists consider balanced reporting (norm of balance) an important feature of journalistic work (Lever-Tracy 2008; Boykoff and Boykoff 2007; Rahmstorf and Schellnhuber 2007; Krauss and Storch 2005). This norm always aims to find two viewpoints to an issue and to give similar importance to both 'sides of a story'. "One might think, that the authority of the IPCC, and the fact that we are dealing with scientific issues, would pretty much dictate what journalists can do in terms of 'spinning a story'. However, there is considerable variance between such alleged facts and their representation to a mass audience" (Grundmann and Krishnamurthy

2010: 128). Storch and Krauss state: "The media, following the U.S. norm of 'balance,' typically present the problem of anthropogenic climate change as a conflict between two opposing schools of thought – and give both schools similar space in advocating their views" (Krauss and Storch 2005: 4). Similarly, Boykoff and Boykoff (2004) discover media biases in a content analysis of reports in The Washington Post, The New York Times, The Los Angeles Times, and The Wall Street Journal; the authors conclude the analysis on a slightly negative tone, claiming that "[t]hrough overwhelmingly 'balanced coverage' of various decisions regarding action due to global warming, the prestige press thereby implied that the division between various calls for action was relatively even. In light of the general agreement in the international scientific community that mandatory and immediate action is needed to combat global warming, U.S. American prestige press coverage has been seriously and systematically deficient" (Krauss and Storch 2005: 134). Rahmstorf and Schellnhuber (2007) find that the picture, the media are painting about the debate is inversely proportional to the development of the actual scientific debate and Trumbo argues that scientists left the debate when public attention increased and were such missing out on the opportunity of getting their message into mass media (Trumbo 1996: 281).

Science Communication: Critiquing the Deficit Model of Information

An expert opinion thus might have the power to get more information – and especially scientific "true" information – into the discourse and thus increase people's knowledge on the issue. But does more information about the issue really lead to people paying more attention to climate change and ultimately change their behavior? At first glance this seems to be an effective path to take in the fight against climate change, since the survey on knowledge of adults (Leiserowitz and Smith 2010) brings about that the less concerned (The Dismissive, the Doubtful, and the Disengaged) also possess less knowledge of climate facts. However, neither Leiserowitz nor Bostrom et al. (1994) and Read et al. (1994) draw the conclusion that more information and knowledge would automatically lead to a higher interest in the debate or even to a change in behavior. There are a few examples in history where this solution has worked, for instance, if we look at the case of cigarette smoking: After a large scientific consensus evolved that cigarettes in fact do harm people's health,[7] a communication campaign started in a lot of countries. Eventually, cigarettes and cigarette smokers were banned from public buildings, school yards, pubs, and restaurants, often enough with public support. Within the last years, social acceptance of smoking declined and fewer young people started smoking (BZgA 2016). When it comes to climate change, however, the case is different (Hoffman 2010). Priddat (2010) argues that the opposite effect would be the case. He takes off from the assumptions of Rawls' "veil of ignorance"[8] and draws the conclusion that the more nations (i.e. governments) know about the impacts and consequences of a changing climate on their own regions, the less they are willing to share the cost with those nations and regions that suffer mostly from the impacts. Even though this study makes an important point it omits that global communities are connected on different levels, like the economy, scientific knowledge exchange or simply the fact of human solidarity. These are aspects that drive the climate change discourse. The role of more and better information on the subject was widely discussed within the realm of science communication, where the deficit model of science communication was first put forward and then widely criticized.

In this section, I will describe the deficit model of science communication even though today its shortcomings are uncovered, because critique of this model is the starting point of what Krimsky (2007) calls the stages in the evolution of risk communication and because it shows how complex the communication of risks and long-term problems is. Risk and science communication underwent a change within the last 30–35 years. Fischhoff (1995), Leiss (1996), and Krimsky (2007) identify three stages of risk communication. The first, the phase of establishing accurate science for the purpose of educating the public, is covered by the information deficit model. The second acknowledges the fact, that simply telling people about a problem is insufficient, and brings the task of persuasion into the communication process. The next step is to change the relationship between communicator and audience from top-down to partnership. Thus, the aim is to engage risk managers, scientists and laypeople in a social learning process (Council of Canadian Academies 2014).

Complex scientific issues such as climate change and related topics pose challenges to "classic" communication strategies, that cannot be met by simply steering the public debate in a scientific adjust direction. Communicating the threads of a changing climate have to deal with invisible causes, with modern humans' disconnection from nature and with delayed or even absent gratification for taking action (Moser and Dilling 2011: 33–34). Scientific information gets transformed within a public discourse, not only through media requirements described above but by being embedded in individual perspectives, values, and views. Thus scientific communication often is at risk of generating boomerang effects, increasing political polarization about highly controversial debated issues instead of leading to a public agreement on scientific consensus (Hart and Nisbet 2011: 6). A lot of work has been put into analyzing public's understanding of science. Bauer et al. (2007) put forward a classification of research paradigms, identifying three groups characterized by research problem and the proposed solutions (Table 2.1).

Here I will limit reviewing this body of literature to what the authors call "science literacy". This paradigm pinpoints to the idea of a public deficit of knowledge about a scientific issue, i.e. a difference between expert opinions and lay people's perception of a complex issue. This model draws researcher's attention to the content of climate change messages, whereas media studies focusing on the norm of balanced reporting (sensational media model) concentrate on structure of communication (Jones 2010: 63). Jones refers to Kellstedt's specification of the model with regards to the issue of climate change. The fundamental assumption of the deficit model of public knowledge is that "scientific assessments of risk are both correct and objective, and then, by implication, the public's perceptions of risks are both inaccurate and subjective" (Kellstedt et al. 2008: 144, cited in Jones 2010: 55–56). Thus, the model promotes increasing information campaigns and getting the "right" information across.

Table 2.1 Paradigms, problems, and proposals [in 25 years of PUS research]

Period	Attribution problems	Proposals research
Science Literacy 1960s onward	Public deficit Knowledge	Literacy measures education
Public understanding After 1985	Public deficit Attitudes Education	Knowledge-attitude Attitude change Image marketing
Science and society 1990s – present	Trust deficit Expert deficit Notions of public Crisis of confidence	Participation Deliberation Angels mediators Impact evaluation

Source: Bauer et al. (2007: 80)

However, various scholars have criticized and rejected the information deficit model (Hart and Nisbet 2011; Bauer et al. 2007; Downing and Ballantyne 2007; Nisbet and Goidel 2007; Nisbet 2005; Gardner and Stern 2002; Goidel et al. 1997). Lowe et al. argue that the approach of a knowledge deficit "failed to place the issues in their wider social and cultural contexts underestimating the depth of public thought and knowledge of risks they face" (Lowe et al. 2006: 435–436),[9] Owens (2000) also emphasizes the importance of framing, when she argues against top-down-communication as inherent to the deficit model. Nisbet and Hart refer to other critics when they reject the model by taking into account "strong value and ideological orientations [that] may act as perceptual screen" (Hart and Nisbet 2011: 3). Studies show that it is not so much insufficient information that prevents taking scientific information seriously or even utter change in behavior, but deeply held values and beliefs, benefits and incentives, social support and peer pressure that play a huge role in foster those changes (Downing and Ballantyne 2007; Gardner and Stern 2002; Semenza et al. 2008; Takahashi 2009; all cited in: Moser and Dilling 2011: 164). Thus, it is urgent to widen the scope of science communication research to understand how audience's attitudes work with the characteristics of informational science messages. Communication research needs to not only take into account mass media as communicators, but consider other sources like social environment consisting of friends, families, workplace affiliates, etc.

The deficit model perceives distribution of information too static when it ignores the fact that people do not just take on an opinion they have read in newspapers and heard in TV- and radio-reports. This is also reflected in Paschen and Ison's (2014) critique of the study by Shanahan et al. (1999), where respondents were asked to select the one narrative out of a variety of constructed environmental narratives, that most reflected their values. This design again establishes the static and hierarchical transmitter-recipient-relationship "by restricting their participants to a passive, receptive role rather than letting them produce their own narratives" (Paschen and Ison 2014: 6). For successful communication strategies social identification and political partisanship are powerful obstacles, which have to be taken into account. In the USA this is especially an issue between Republican and Democratic views. Al Gore's popular 2006 documentary on global warming "An Inconvenient Truth" and the 2017 follow-up "An inconvenient sequel: truth to power" turned the issue almost completely into a Democratic topic. Political elites, whose ideological orientations tend to be stronger than those of the general public, show a huge politically fed divide when it comes to existence, causes of, and solutions to climate change (Hart and Nisbet 2011; Dunlap and McCright 2008). The strong polarization among political beliefs stains public opinion, especially considering the highly emotional presidential election campaigns in the USA. "As beliefs about climate change become strongly associated with partisan orientations individuals are more likely to pay attention to and interpret information in ways that reinforce their political views" (Hart and Nisbet 2011: 2). A finding, that is seconded by the current debate about discursive bubbles and filters, especially in social media, that reinforce people's worldviews and that block out different perspectives and viewpoints. Ideological polarization prevents an objective scientific debate, more than that: it shows in a nutshell that problems, no matter how 'scientifically objective' they may seem are not negotiated in a vacuum, but are embedded in issue cultures and bridging metaphors (Ungar 2007: 81). Thus, putting culture into the central assumption of communication analysis can help us understand the background noise (Norgaard 2011) of climate change debate.

Results of these studies are important to understanding expert communication, because it is safe to assume that experts are aware of media requirements and consequential challenges for communicating climate change. If they want to be heard they have to obey these rules.

The Role of Culture in Climate Change Research

One paradigm coming from the realm of humanities and research interested in the role of culture in our social life is to understand the relation between causes and responses to climate change and culture. This is one of the core competencies humanities and social sciences can contribute to the debate about climate and environmentalism: re-defining and re-evaluating the dialectical relationship between nature and humans. Questions and issues, this strand of research addresses are:

- The socio-cultural interpretation of climate change as a social phenomenon (e.g. Leggewie and Welzer 2010; Heidbrink 2010)
- Diagnosis of the culture-nature-relationship (e.g. Chakrabarty 2010; Priddat 2010)

- Reflections on globalization and herein global justice as it is presented in questions of climate justice between nations and even continents (e.g. Clausen 2010; Messner 2010)
- Social-ecological explanation of interdependencies between natural and cultural processes (Becker and Jahn 2006; Fischer-Kowalski and Weisz 2005).

Environmentalism has long left the niche where it is "just" about nature itself but managed to draw direct connections to fundamental social issues like human rights, social justice,[10] and political economies (Eyerman and Jamison 1991). This relationship between nature and culture is the main interest in social ecology. In fact, the modern categorical differentiation between nature and culture is a fundamental precondition for the development of social ecology in the 1980s (Becker and Jahn 2006: 29). However, social ecology aims at avoiding a naturalistic or culturalistic conception of society-nature interaction (Fischer-Kowalski and Weisz 2005: 114). Based on this, social ecology describes social-ecological phenomena by analyzing interdependencies between social and natural processes (Becker and Jahn 2006: 189). "We assume that culture (conceived as systems of recursive human communications) and nature (conceived as systems of the material realm) are dichotomous, and we attempt to construct a kind of interface between these two realms that is capable of explicating interactions." (Fischer-Kowalski and Weisz 2005: 135). Figure 2.3 shows

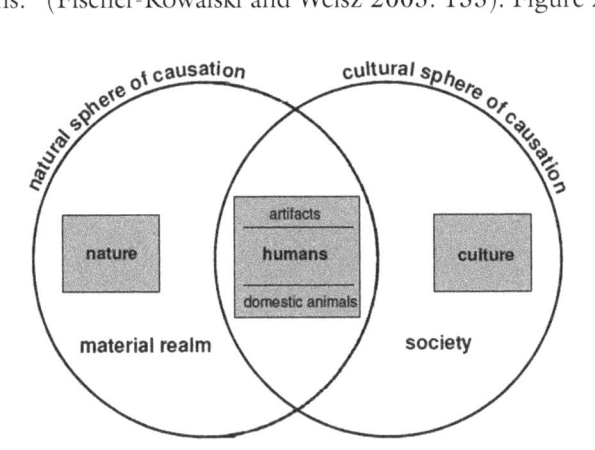

Fig. 2.3 Social ecology – the interaction between cultural and natural spheres of causation (Source: Fischer-Kowalski and Weisz 2005: 137)

the interaction between cultural and natural sphere with humans as communicators and living beings at the center of the intersection. Here, communication is influencing the cultural sphere of causation, which in turn is constituted by culture; the natural sphere encompasses the whole of the material world.

This interaction sets out to fill the epistemological gap between society and nature, identified by scholars within the realm of social ecology. Drawing on ecology, Fischer-Kowalski (1997) conceptualizes the nature-society interaction as societal metabolism, a term that has been used before in Marx and Engels' description of the labor process, as well as in other contexts in the realm of social geography and cultural and ecological anthropology.

The use of biological terms as applied to social phenomena is justifiably still a controversial topic; however, the concepts approach to nature-society-interaction brings the idea of anthropogenic influence on nature into the debate. Entering the debate about climate change, this development is taking a step further when a system as complex and seemingly robust is altered by human influence. Chakrabarty (2010) even calls this age the "anthropocene", that is, an era in the history of the world, where human influences on ecological developments are highly significant. Climate change means that boundaries of ecological imperialism that characterized especially the industrial revolution are reached and that human superiority is challenged by revolting nature. That twists human history to the end of the theological creation myth according to which man was called to govern nature and all ecological things. Priddat (2010) argues that the notion of an anthropogenic climate change keeps up this interpretation of the nature-men-relationship ex negativo without acknowledging the active role of nature. Human history was never thought as environmental pawn and against better judgment climate politics are undertaken in terms of political bargaining about safeguarding of national interests and exploring divergences. The debate about climate change itself and fundamental semantic meaning of terms, used in the debate thus becomes subject to knowledge-theoretical reflection (Daschkeit and Dombrowsky 2010; Büscher and Japp 2010). Büscher (2010) claims the ecological crisis to be more a self-endangerment to the human race than an endangerment to earth and nature, and that a philosophical framing has to emphasize this. Not only the relation between ecological sphere and anthropogenic sphere are widely discussed, but also the idea of a global climate change is questioned. Clausen (2010) identifies three dimensions – rapidity, radicalism, and rituality – along which he analyzes the development and intersection between local, global, and glocal climate situation, arguing against a "global climate catastrophe", since worldly developments are not intertwined in all three dimensions. Priddat makes a similar point when he describes territorial instability as consequence of regional differentiation. Politics under the impression of climate change do not lead to a strong clannishness but on the contrary to decentralization because of varying impacts of climate change in different parts of the world. Clausen suggests using the term of multiple cultural catastrophes, not least because in the end, every cultural sphere has to fight its own climate crisis within its own possibilities and abilities. The globalization of global warming is also subject in Messner's (2010) reflections about the discourse itself, where he states three steps in development: the first globalization discourse (globalization 1.0) gave rise to dissolution of economic boundaries, leading to an economic globalized world. The second step, globalization 2.0, leads to shifting power-relations. Finally, globalization 3.0 is fed from a debate of climate and development, centering around globally significant but locally limited tipping points.

Leggewie and Welzer make a central claim that climate change with regards to its consequences ought to be subject of social sciences and cultural studies (Leggewie and Welzer 2010: 31–33). The authors depict global environmental changes as rooted in a global cultural; therefore, a cultural change is needed if climate change shall be tackled sustainably. Leggewie and Welzer argue that this will be the biggest challenge we will be facing, since habitual and cultural courses are difficult to reflect about. The ability not to question our lifestyle on a daily basis is what keeps a society going, but is also the reason, why a necessary cultural change will be difficult to achieve. Culturally determined values, norms, and visions of life stand in the way of changing societal habits (Minkmar 2010). Evermore, Heidbrink points out that climate change is even a consequence of a modern cultural understanding of nature as being subject to rational disposability,[11] thus climate change is to be treated foremost as a cultural project (Heidbrink 2010: 52). As such, climate advocates need to take into account that global change deals with culturally different perception patterns. These shifting baselines go for timely changes, when following generations have a different mental image of nature than generations before and also goes for differences in the ways people are used to deal with nature (Leggewie and Welzer 2010: 35).

From this short review it becomes evident that the debate needs a humanistic perspective that takes into account cultural challenges that come with the territory (Minkmar 2010). An examination of cultural stories about climate change and the fight against it will contribute to this approach. Here, Norgaard (2011) identifies a need for a macro perspective when analyzing people's response to climate change:

> Most existing studies on public response to climate change – coming from environmental sociology, social psychology, or science communication, from survey works on attitudes and beliefs to psychological studies on mental models – use individuals as their unit of analysis. Yet [...] social context itself can be a significant part of what makes it difficult to respond to climate change. Although studies from this literature are essential, studies of perception that focus solely on individuals are unable to grasp the meaning of differences across cultures, subcultures, nationality, or the influence of political economic context on how individuals and communities think, feel, and imagine. (Norgaard 2011: 209)

Cultural Studies and Cultural Theory Approaches

The following literature I am presenting here deals with the cultural implication of the climate change phenomenon, when the discourse is understood as a set of stories or, as Daniels and Endfield (2009) put it: a Big Story, that bring different aspects of the problem into the debate. The following studies prominently deal with questions like

- Which role does culture play in the perception of climate change? (e.g. Leiserowitz et al. 2010; Osbaldiston 2010; Krauss and Storch 2005)
- What is the cultural and social background of climate change stories? (e.g. Norgaard 2011; Ney and Thompson 2000)
- What are the characteristics of climate change narratives? (e.g. Smith 2012; Jones 2010; Verweij et al. 2006; Smith and West 1996)

With these questions, the respective authors open up the possibility of reading climate change as social text and set the stage for an analysis of its cultural structures. We have already seen the importance of communication, now it is time to understand the consistency of single communication elements.

Shanahan et al. argue, that "narratives are important to an understanding of environmental issues. [...] Narratives are a separate and distinct type of communication that people use to make decisions about environmentally relevant issues" (Shanahan et al. 1999: 408). The authors conducted a mail survey, which featured different narratives on an environmental issue, but ended up with inconclusive findings in this explorative study. However, the results show "that narrative assessment of environmental concern should receive further attention [... since] the use of narrative techniques improves the predictiveness of attitude-behavior models" (Shanahan et al. 1999: 416–417).

In her study "Living in denial" Norgaard (2011) puts forward an ethnographic approach to understanding culturally and socially bound perception of climate change impacts in a small, fictional Norwegian town she calls Bygdaby. The study stands out from the body of climate change literature because it does not use an instrumental approach but in a Geertzian sense of thick description digs deep into socially and culturally constructed (collective) emotions towards ecological changes, and thus commences in this review the transition towards a cultural sociological view on climate change. Norgaard traces cultural strategies to dissolve the dissonance between ecological awareness and supporting national economic policy, which is especially important to a country like Norway where being in touch with nature is part of the national identity (Norgaard 2011: 146–151). On the center of the study stands the idea of "denial work", that is the social effort that is put into denying the consequences of climate change not only on a personal level, but even more importantly on a collective level. "Ignoring" a threat thus is not passive, but becomes active work, conducted collectively. Non response to climate change is produced through cultural practices of everyday life, allowing for normalization by use of cultural narratives: "Various narratives, most either produced or reinforced by the national government and echoed by citizens, serve to legitimize and normalize Norwegian climate and petroleum policy" (Norgaard 2011: 140) and to minimize Norway's responsibility for the problem of climate change. Norgaard identifies three categories: the narrative "Norway is a little land" and the narrative "We have suffered" both refer to Norwegian identity and history, describing the nation's relatively small contribution to global CO_2 emissions because (1) of the small population and (2) of its economic history, that Norway has – until somewhat recently – always been a poor country. The third narrative "America as a tension point" stirs the debate away from Norway itself and brings in additional characters as projection surface by pointing out other nation's bigger faults (Norgaard 2011: 142). Norgaard's description of Norwegian discourse is in line with studies on attitudes and beliefs stating that people are in fact concerned about climate change and have knowledge available, but at the same time "don't really want to know and in some sense don't know *how* to know" (Norgaard 2011: 207; italics original). But it takes the question of non-response to climate change a significant step further by taking into account cultural constructions of collective emotions when examining why people do not care (enough) about climate change and change behavior patterns accordingly. Although studies have considered the role culture plays in the perception of issues like climate change (Krauss and Storch 2005) they yet focused on the cultural impacts of individual perspectives, using, as cited in Norgaard above, the individual as unit of analysis. Norgaard's approach and the studies I will introduce in this section, however, assign culture the role of a gateway for the examination of climate change discourse, thus acknowledging the texture of societies.

Placing critique on the norm of balanced reporting and the deficit model of information as explanation of laypeople's lack of awareness and lack of willingness to fight climate change, Jones argues that "individuals do not process information in a vacuum; rather, individuals bring their life experiences and their understanding of the world to bear when determining what information to accept, what information to reject, and, most importantly, what that information means" (Jones 2010: 64). Bringing the idea of culture as action- and value forming concept into the climate change debate, scholars have turned to Cultural Theory, as it was prominently laid out by anthropologists Mary Douglas and Aaron Wildavsky (1983, c1982) in their study on cultural risk perception and in Douglas' analysis of the social meaning of dirt (Douglas [1966] 2004), thus showing how essential social history and cultural context is for examining everyday life practices. With her work on pollution, Douglas already hints at the importance of cultural beliefs in social organizations, which is further explored in the grid/ group model. The model takes into account that societies are based on systems of classification (Smith 2001: 83). Douglas argues that the perception of purity vs. pollution can only be understood with a consideration of the wider classification that is at work in a society. "Things, that did not fit into orthodox classification systems and which violated or crossed symbolic borders tended to be seen as polluted" (Smith 2001: 83). These observations inform the four-cell-typology that explores the importance of systems of classification for social organization (Fig. 2.4).

Four Rationalities

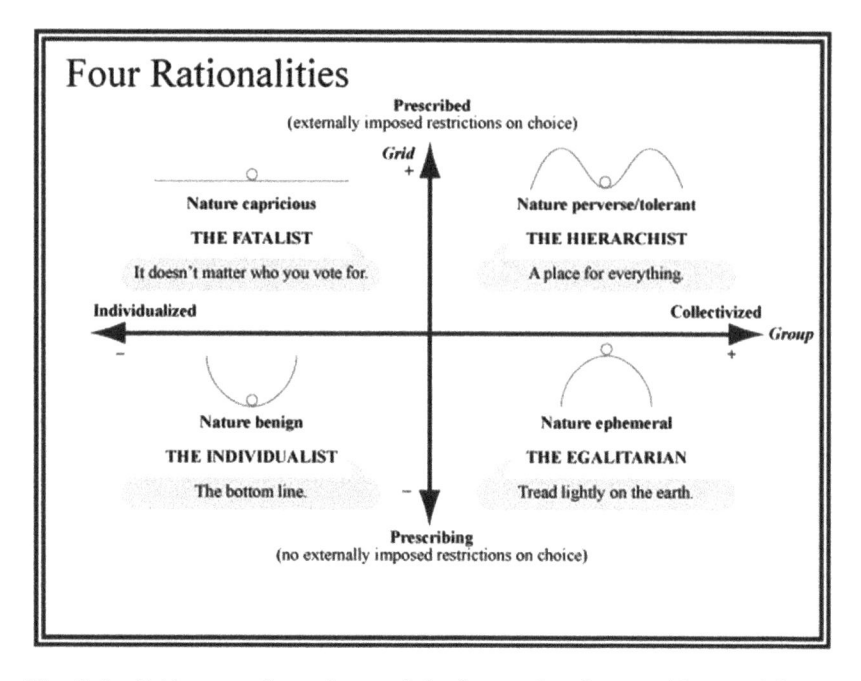

Prescribed
(externally imposed restrictions on choice)

	Grid +	
Nature capricious		Nature perverse/tolerant
THE FATALIST		THE HIERARCHIST
It doesn't matter who you vote for.		A place for everything.
Individualized −		**Collectivized** + → Group
Nature benign		Nature ephemeral
THE INDIVIDUALIST		THE EGALITARIAN
The bottom line.	−	Tread lightly on the earth.

Prescribing
(no externally imposed restrictions on choice)

Fig. 2.4 Grid-group dimensions and the five myths of nature (Source: Schwarz and Thompson 1990: 9)

Within Cultural Theory there are four ideal types of value orientation created, along the two dimensions of group and grid. Chai et al. summarize this analytic model: "The Douglas model proposes that an individual's behavior, perception, attitudes, beliefs, and values are shaped, regulated, and controlled by constraints that can be categorized into two domains: group-commitment and grid-control" (Chai et al. 2009: 195). Grid/group analysis is "a way of checking characteristics of social organization with features of the beliefs and values of the people who are keeping the form of organization alive. Group means the outside boundary that people have erected between themselves and the outside world. Grid means all the other social distinctions and delegations of authority that they use to limit how people behave to one another" (Douglas and Wildavsky 1983, c1982; italics original). "Grid measures preferred levels of group interaction, while the dimension of group captures the degree that these groups are expected to constrain the individual's beliefs and behavior"

(Jones 2010: 67). Cultural Theory allows placing individuals along the dimensions of grid and group, with 'grid' as a dimension of individuation and 'group' as a dimension of social incorporation (Douglas 1982: 190), creating five types: fatalist, hierarchist, individualist, egalitarian, and hermit. Policy studies focus mostly on the first four types, arguing that the hermit's worldview does not have consequences for social life (Thompson et al. 1990). The four types then can be assigned a specific perspective on nature.

The four myths of nature serve as simplest models of ecosystems, partially representing reality (Thompson et al. 1990: 26). The individualist in the grid/ group fourfold table sees nature as friendly and harmless (myth: nature benign). The world is perceived as forgiving, allowing for trial and error strategy when faced with uncertainty. The egalitarian's myth of nature (nature ephemeral) can be seen as the opposite to the individualist's perspective. Here, nature is perceived as unforgiving and terrifying, and can be brought out of balance with smallest steps. Thus, societies have to treat nature with great care and timorous forbearance. The myth of nature perverse / tolerant encourages the hierarchist to be suspicious of unusual occurrences, since nature is only forgivable to some extent. Thus, societies have to ensure that exuberant behavior towards nature and its resources does not exceed a certain level. The fatalist sees nature and the world as random, leading to the attitude that societies are best off by treating unusual occurrences as singular events. Here, social managing institutions do not learn and progress (and do not have to), whereas the myths of nature benign, nature ephemeral, and nature perverse/tolerant require learning process when coping with environmental crises (ibid.: 26–38).

Leiserowitz (2003) and Jones (2010) take cultural theory as basis for their studies on climate change opinions among U.S.-Americans. Both studies aim at operationalizing cultural theory to "test the theorized relationship between cultural worldviews and risk perception of global warming" (Leiserowitz 2003: 53) resp. to "operationalize [...] narrative theory [...] seeking to determine if cultural narratives help explain variations in climate change opinion related dependent variables" (Jones 2010: XI). In a critical reflection on cultural theory, Leiserowitz criticizes Douglas' focus on the strictly causal relationship between social relations and cultural worldviews: "While Douglas argued that social relations determine worldviews (strong social constructivism), most subsequent scholars argue instead that the two are dialectically related" (Leiserowitz 2003: 46). Leiserowitz concludes – in line with the critique of communication research as aiming at "one size fits all" – that multiple strategies are needed for successful climate change communication, relating to different groups of audience, those who are skeptic[12] about the existence and causes of global warming or those who lack of scientific understanding, as well as to different implications of climate change, like climate change as a threat to human health issues which did not show as especially troublesome to respondents. Based on successfully constructed indices for the worldviews of egalitarianism and fatalism the study supports Cultural Theory's prediction that egalitarians are more concerned about environmental risks, that climate change activists would be significantly more egalitarian than the average of the U.S. American public, and that fatalists' risk perceptions, policy preferences or individual behaviors do not vary significantly from the mean. Indices for hierarchical and individualistic worldview however did not show internal consistency (Leiserowitz 2003: 191–192). The fourfold typology of forms of social solidarity emerging from cultural theory were in recent scholarship combined with the study of narratives – or stories – which lead to understanding synergies between both approaches.

Ney and Thompson (2000) point out three dominant stories about global climate change[13]:

Profligacy: The first story centers around "profligate consumption and production patterns in the North as the fundamental cause of global climate change" (Ney and Thompson 2000: 71) and is part of an egalitarian setting. This story embeds the case of climate change into a wider societal context of the logic of profit motif and focus on economic growth, with the villain found in exactly these structures of industrialized countries. On the other site, those organizations that understand how humans and nature are linked and that can see beyond short term satisfaction in western capitalist culture are the heroes in this tale. The moral of the story thus lies in a change of culture, where we "move from the idea of a waste society to the concept of a conserving society" (Ney and Thompson 2000: 73).

Price: Ney and Thompson head the second big story line with the term "prices". This economic tale locates climate change in the realm of technical solutions and makes natural resources subject to market forces. Resulting here is a technical discourse about nature instead of a debate about comprehensive cultural change. However, that does not exclude a discourse about a sustainable economy. In order to achieve such a green economy sufficient funds will be necessary which again can only emerge from economic growth. In this story, rational individuals are capable of making their own decision, locating this story within the idea of individualism, misguided policy makers, which are meddling with market forces, are found to be the villains of the story, whereas those organizations which understand the economics of resource consumption are the heroes. The moral can be summed up in the demand "we have to get the prices right" (Ney and Thompson 2000: 74), meaning that the market will take care of reducing consumption of natural resources with the help of financial (negative or positive) incentives.

Population: The third big tale takes place in a hierarchic setting and is described by the tag "population", which pretty much pinpoints the core of the story: uncontrolled population growth necessarily leads to an over-consumption of natural resources, and is thus identified as the villain. As a consequence the moral of the story is to rationally control global population growth. The heroes in turn are those actors that are equipped with the organizational capacities and moral responsibility to tackle the problem. Thus the combat against climate change should be left in the hands of experts.

Based on this previous work, Verweij et al. (2006) also follow the three (four) ideal types of social solidarity when turning back to Ney and Thompson's classification.

Profligacy – an egalitarian story: In this story, earth is seen as vulnerable to over-consumption (setting), a villain is easily identified in industrialized countries whose consumerist lifestyle makes use of natural resources with no or little regard to the consequences nature and people in poorer regions of the world have to suffer. Heroes in this story are activists and NGOs. Climate change is thus perceived as a morally and ethical issue (Verweij et al. 2006: 822).

Lack of global planning – a hierarchic story: In this view climate change is seen as a serious thread, however, it emphasizes the long-term characteristic of the issue, meaning that there is still time to provide solutions, placing climate change within a setting of a tragedy of the global commons. The alarmist notion, which is embedded in the first story, is missing here. NGOs, scientists, and politicians who sustainably work towards a non-carbon age at a well-wrought and responsible paste are the heroes in this story. Those individuals and organizations that do not subscribe to intergovernmental treaties to meet climate change are thus the villains in this story (Verweij et al. 2006: 824).

Business as usual – an individualistic story: Supporter of this view see climate change as a scare made up by international bureaucrats and scientists trying to expand influence and secure research funding. The eco-system is understood as robust to human influences (setting), decision makers who do not act as puppets to this scare are heroes while those are the villains that feed the myth of a soon destroyed earth and scare the public. Climate change, if it exists, can be met with clean technical and innovative solutions (Verweij et al. 2006 825).[14]

These three stories, conflicting and clashing as they are, represent one of the biggest cultural impediments in dealing with climate change: they are not reducible to one another, and as immanent value orientations, cannot be proven wrong or right (Thompson 2003). In other words: "[…] people are arguing from different premises and […] since these premises are anchored in alternative forms of organizing, they will never agree" (Verweij et al. 2006: 821). Thus, climate change will always be subject to at least three different stories any policy process has to deal with that by taking into account each segment that is singled out in every story. Keeping in mind that these stories are based on ideal types that are lacking analytic purity in real life it is worth looking into other stories and narratives that surround climate change. Pursuing this approach with narratives and narrative structures in the center I will give a short overview of further studies examining narratives of climate change.

Studies that examine the way climate change is culturally treated in social debate are discordant when it comes to the proper term that should be used. This depends of course on the author's scientific school and its perspective, but overall it seems like most of the terms are interchangeable in the realm of this study (Table 2.2):

Heyd (2010) for example speaks of cultural frameworks when he describes societal dominant prerequisites that influence the perception of climate change.[15] However, he also starts from cultural theory paradigms, as do the above presented studies which make use of the terms story and narrative. At the same time, we find Liverman's analysis (2009) focusing on the term of narratives, but in a broad social scientific sense, lacking of narrative theory paradigms, such as identifying specific narrative elements. Hence the depicted narratives rather resemble characteristics of discourse sequences. Especially in Verweij's et al. (2006) study we can see the results of the so-called narrative turn in social sciences research, revealing narrative structures within social dispute and thus offering tools for discourse

Table 2.2 Overview: terminology and topics in climate change communication literature

Author	Terminology	Key findings
Norgaard (2011)	Stories	3 stories supporting climate change denial (Norway is a little land; We have suffered; America as a tension point)
Jones (2010)	Narratives	Cultural narratives as explanation for variations in climate change opinion
Ney and Thompson (2000)	Stories	Egalitarian story of profligacy Individualistic story of price Hierarchical story of population
Verweij et al. (2006)	Stories	Egalitarian story Hierarchic story Individualistic story
Liverman (2009)	Narratives	Narrative of climate change as investment opportunity
Trumbo and Shanahan (2000)	Narratives	Symbols in narratives: polar bear as climate change symbol
Bravo (2009)	Narratives	Narrative of a victimized community (Arctic identity)
Osbaldiston (2010)	Narratives	Unlocking knowledge how myths, symbols influence perception of climate change
Smith (2012)	Narratives	Apocalyptic narrative of climate change
Smith and West (1996, 1997)	Narratives	People against nature narrative: droughts as outside enemy in Australian society
Bettini (2012)	Narratives and discourse	Doom and gloom narrative of climate refugees: humanitarian or national security agenda
Daniels and Endfield (2009)	Stories	One big story
Ereaut and Segnit (2006)	Frames and repertoires	Alarmist repertoire Optimistic repertoire Pragmatic repertoire
Hamblyn (2009)	Narratives	Climate change developed from environmental news story to a narrative of human responsibility
Leiserowitz et al. (2010)	Frame	Climate change developed from an environmental problem to a political problem to a human/public health issue
Shanahan et al. (1999)	Stories	Climate change developed from a scientific problem to a political problem

(*continued*)

Table 2.2 (continued)

Author	Terminology	Key findings
Michaelis (2000)	Narratives	Modernism: climate change as optimization problem with a technical solution Romanticism: climate change as problem of the modernist attitude (consumerism, etc.) European humanism: climate change is a result from the failure to nurture virtues like humility and continence
Moser and Dilling (2007)	Communication	Rejecting the fear approach to climate change communication
Myers et al. (2012)	Frame	A public health and national security frame for climate change
Marshall (2014)[a]	Narrative	Identity markers in narratives for successful climate change communication

Source: Own illustration

[a]Recently, research on narratives with regards to climate change has found its way out of the purely academic interest and into applied studies. The UK-based advocacy group Climate Outreach (www.climate-outreach.org) managed an event where narratives were tested in discussion groups with laypeople (Marshall 2014). In this applied study, narratives were used in terms of content or topics, centering around one specific issue like national Welsh identity.

analysis. Studies dealing with narratives feature different empirical approaches, from looking into newspaper articles (both as representations of the social debate as well as journalistic influence), to using large-scale survey data on climate change opinion and knowledge through in-depth interviews as focus groups. The turn towards narrative analysis offers new tools to examine climate change as s social phenomenon, "narratives have a natural dramatic arc to them, and thus may predict how audience's attention for stories will wax and wane" (Trumbo and Shanahan 2000: 201), it offers an innovative approach into the old question what is missing from the climate change story that will capture people's attention. Not only media coverage of the issue entails a translation of scientific facts into meaningful and easy-to-grasp stories by orchestrating facts around a dramatic structure, including a storyline, characters, a setting, and a theme,[16] but also social debate, consisting of written, oral, and visual text, works that way. Trumbo and Shanahan argue for such a narrative turn in climate change discourse analysis when they state: "If public understanding of this issue is built on a narrative construct, then policy and regulatory strategies that rely on an authority located in public opinion could be seriously misinformed" (Trumbo and Shanahan 2000: 203). Liverman argues similarly that an analysis of narratives of the climate change discourse might reveal ineffective and unequal policy strategies based on obscured historical geographies. "The narratives have been employed to design an international response to climate change that has been influenced by powerful political interests and has embraced the neo-liberal project of market environmentalism" (Liverman 2009: 280). The author names key sequences within climate change history: one refers to the manifestation of a dangerous climate change in the earlier days of climate change discovery; another one describes the controversy about human influence on climate. This contains an answer to the question of allocating responsibility and blame, ranging from a global collective to nation states through companies and individuals. International treaties however are based on nation states. This in turn leads to the third sequence, which emphasizes market solutions to climate change, such as carbon trading (Liverman 2009: 287, 288, 292–295). Drawing from these findings Liverman makes out a newly emerging direction of climate change as an investment opportunity, putting climate change into the realm of market environmentalism.

While the above studies look into the evolution of narratives and their social and cultural setting, Bravo (2009) focuses in his study on people's reception of narratives, more precisely on Arctic citizen's reception of a dominant climate change narrative that orchestrates the relationship between industrialized and developing countries with respect to climate change impacts. Although the study of the analysis lacks proper analytic systematization, it stands out from the realm of climate change narratives because firstly, it focuses on the reception and secondly, emphasizes an aspect that is often neglected in cultural studies but emerges more and more in development studies: the perspective of industrialized countries on poorer regions in the world, attributing a dominant role to western societies. In the case of the Arctic that Bravo is examining, the narrative of the environmental crisis produces "a new Arctic regional identity in which citizens, particularly indigenous groups, are simultaneously portrayed as being an 'at-risk community' [and] a victimized community lacking the agency to fight back" (Bravo 2009: 257–258). The study thus inherently inquires after the relationship between actors that is presented in different narratives, even though this question here is covered underneath demands of development studies.

The cultural theory approach allows for culture to take a major role in the drama of global climate change. Scholars apply cultural studies to the phenomenon and hence create a cultural field for policy implications. But climate change is not understood as a cultural phenomenon in terms that a changing global environment is not a problem in itself, but becomes a problem only through cultural lenses. For this I will at last introduce climate change and environmental studies in the realm of cultural sociology.

CULTURAL SOCIOLOGY APPROACH TOWARDS CLIMATE CHANGE

The evolution of narratives shows that natural sciences only deliver raw facts, which in turn are in need of social and cultural interpretation to become social facts (Smith 2012: 758; Hoffman 2010: 295–296). That is what some representatives of the humanities and social sciences point out, when they talk about the interpretational sovereignty that the social and cultural sciences need to claim back (Leggewie and Welzer 2010; Welzer et al. 2010; Hagner 2010). Scientific findings about the development of global climate do not offer a dogma or axiomatic truth but merely deliver information bits and pieces that are subject to societal discourse. Studies described above offer valuable insights when it comes to understanding, how culturally transformed communication within society portrays climate change. Media logic, communication challenges of climate change as time-and-spatio-distant phenomenon, and the need for narration of scientific facts have all been presented as strategic tools to a solution for successful climate change communication (e.g. Hart and Nisbet 2011; Jones 2011; O'Neill and Nicholson-Cole 2009; Moser and Dilling 2007; Downing and Ballantyne 2007). However, climate change is still seen as something outside the social and cultural realm which in fact has to be dealt with in a manner that is processible for our social way of thinking and acting, but it still does not understand climate change itself as cultural. It is looked at as a fact from natural sciences and thus per se as outside the cultural system. Yet, cultural studies and approaches from the realm of sociology of culture miss the opportunity to acknowledge natural sciences – and their products – as part of the socio-cultural system. Yes, it is true that climate exists with or without human action, but its meaning is a priori culturally embedded. In other words: climate change as social phenomenon, no matter how 'natural' it is, is spun into a deep web of meaning and constitutes as such a cultural and social object. This is where a cultural sociological perspective comes into play. Climate change has to be considered as "a meaningful social fact, [... it is] signifier and drama in a surprisingly complex cultural field" (Smith 2012: 745). Before laying out the basic assumptions of cultural sociology and its advantages in understanding social communication about climate change, I will present statements developed in studies coming from a cultural sociological perspective.

Sociologists and other representatives of the 'soft' sciences often settle for the role of interpreters of quantitative surveys, "data mongers, and policy critiques" (Osbaldiston 2010: 7), or are pushed into that role in the dialogue with 'hard' sciences. Contrary to that, Osbaldiston sees cultural sociology in a supporting role for the development of climate change adaptation and –mitigation policies and frameworks. He points out that "cultural sociology holds the key to unlocking knowledge on how myth, narrative, and public discourse [...] interact with 'hard' data and dominant paradigms" (Osbaldiston 2010: 7). Such a cultural perspective might even widen our understanding of the relationship between knowing and acting on that knowledge, as is shown in the value action gap and the critique on the knowledge deficit model. Myth and symbolism influence just as much people's course of action as cognitive information do; that is by serving as a filter for raw scientific data. This provides a cultural interpretation of the fact that different nations, different communities, NGOs, and politicians react in very divergent ways to the same basic information about global warming (Smith 2012). Considering questions about people's beliefs and attitudes towards climate change, cultural sociology can help to understand how scientific knowledge interacts with cultural discourses. Knowledge and insights into the working of social groups and subgroups and cultural structures of communication enables cultural sociology to provide deep findings in this area, fruitful for subsequent policy strategies. Those findings are not only drawn from data on climate change discourse but from putting well known paradigms of sociological theory to use in understanding modern communication about such a contemporary phenomenon. This can be seen in Smith's essay on narrating global warming (Smith 2012) as well as in Smith and West's study on public discourse about droughts in Australia (Smith and West 1996, 1997), where Durkheim's paradigm of social solidarity rooted in collective representations is seen in public disaster text and talk. Drought is presented as the outside enemy that is threatening the social group. "Drought, then, is the alien force against which society must unite" (Smith and West 1996:

95). This cultural construction of nature appears in coding each drought as unprecedented and special, even though drought years are a fairly regular phenomenon in Australian (weather-) history. Smith and West see a consistency with Durkheimian functional theory, "which predicts that collective representations will sometimes outstrip reality in the interests of social solidarity", with droughts operating as a signifier in a 'people against nature' narrative (Smith and West 1996: 95). Australian drought discourse thus can be understood as moral drama in Durkheimian terms, setting itself apart from standard moral panics literature which "suggests that most national cultures make use of human enemies as scapegoats in such solidarity-engendering narratives" (Smith and West 1996: 97). However, natural disaster events epitomize several characteristics suitable for this cause. For once, natural disasters cannot initiate a counter narrative; they are, as Smith and West put it: mute. As such, they display a phantom objectivity, i.e. they happen beyond party politics. People perceive droughts as something that really happens. Even though this perception is changing over time, especially among agricultural elites and environmentalists, disasters like cyclone, floods, earthquakes etc. are still considered to be more closely connected to human agency. Devastating consequences from an earthquake, like collapsed buildings, can be traced back to insufficient construction, floods can be hold back by proper dam building and flood protection measures; opposite to that droughts in the Australian mind just happen and have to be dealt with. This perception can be understood with reference to Australian mythology where droughts have always played an important role. This is one of three key variables that distinguish droughts from other natural disasters (Smith and West 1997):

- *The variable of mythology*: the specific climate and characteristics of landscape have always played a significant role in Australia's identity construction, especially to contrast the nation from Europe and Britain. Contemporary discourse about droughts thus can make recourse to tales anchored in the nation's history which allows newspaper articles and political speeches to expect common understanding of droughts from the audience.
- *The variable of time*: unlike other sudden and unprecedented natural disasters droughts last a significant duration, and thus provide the opportunity for public discourse to construct them as national enemy. Its frequent occurrence assures that debate does not completely vanish; on the contrary it allows for a coherent and ongoing tale.

- *The variable of space*: droughts are widespread events, stretching state boundaries and different regions, thus, they fit for a national narrative.

There are a number of analytic findings from Smith's and West's study that can be applied to the case of climate change, even though the study is not entirely pertinent. For example, it is safe to say, that global warming is not 'mute', but rather a highly politicized issue with many different actors raising their voices in the debate. Also, global warming is not a frequently occurring phenomenon, but an ongoing process over time with more or less visible consequences to the date. The resulting events like floods etc. are timely limited; the process itself however is not. On the other hand, global warming is per se a widespread – indeed global – process in need of global efforts and thus stressing the variable of space. Most important in this study is how the authors embrace the perception of natural disasters with a cultural approach by investigating the socio-cultural meaning of droughts. Similarly, Smith analyses the ongoing debate about climate change in cultural terms (Smith 2012). The drama of climate change is analyzed according to its narrative structures, referring to the Structural Model of Genre (Smith 2005: 24), which displays different genres – ranging from low-mimetic to tragic, romantic, and apocalyptic – in combination with a variable of characteristics, featured within a specific genre. Similar to the critiques of the information deficit model, Smith declines the assumption that the more scientific facts are known, the more societies start to worry and take action. He points out that knowledge about the chemical processes of global warming are merely one side of the story. Modern societies are still influenced by pre-modern doubts and anxieties, wrapped in myths and narratives. Scientific information thus provides the basis for a new societal narrative, but before it becomes relevant to a wider public (that is outside the scientific circle) it needs to be transformed into another set of codes, it has to become storied (Smith 2012: 746). The story of climate change has – according to Smith's analysis – undergone a genre shift throughout the last few years. During the 1980s, little attention has been paid to the subject, due to a lack of assured facts. As long as knowledge on the matter is still vague, there seems to be no need for immediate political action. Smith categorizes this stage as a romantic minimization of the problem.

As newscasts and newspapers picked up the notion of a changing global climate, scientific concepts became familiar to laypeople, using formerly exclusive scientific concepts in everyday conversation and thus getting engaged in the whole debate. The topic is no longer set in a scientific context but becomes politicized and socialized, "news work is not merely an instrumental task of filling the news hole. […] Instead, news work requires the transformation of discrete events into meaningful narratives" (Jacobs 1996: 392–393). Reflecting the growing influence of environmentalism, an apocalyptic genre guess took over the debate, putting huge pressure on the 1997 UN meeting in Kyoto and producing high expectations for its outcomes. The rise of an apocalyptic genre can also be seen in the rise of climate change symbolism. For years now, the image of a drowning polar bear – prominently displayed in Al Gore's environmental documentary – stands as a pars pro toto for those devastating consequences a changing climate is likely to bring and that fuels environmentalist's demand for immediate political action. To keep a story in a collective memory, it needs a symbol that people can associate with the more complex, wider background (Espeland and Micheals 2012; Trumbo and Shanahan 2000: 202). Cultural structures are not only textual, but are supported and strengthened by symbols and signs (Alexander 2003: 24). For climate change, the polar bear has become the picture, and the 2°C[17] target, that was so vividly proclaimed by scientists, policy makers, and environmental activists, became the significant simple number. It is a telling example how scientific coherences get illustrated by symbols in order to carry understandable meaning. When the UNFCCC[18] agreed in 2010 to keep the target of 2 °C, the number that was first put forward by economist William Nordhaus in the 1970s; the number gained new media attention, evolving as a symbol a wider public can recognize. Media coverage of the 2010 UNFCCC introduced the 2 °C as a symbolic figure to the public debate on global warming. As much as the polar bear it became more than just a number, it became a collective representation. Similar transformations can be seen with the Occupy Wallstreet Movement in 2012, where the number of 99% has become a shared symbol for everyone engaging in the protest ("we are the 99%"), just like the 10% figure has been used for illustrative purposes within the modern gay rights movement and gay identity politics (Espeland and Micheals 2012). Symbols – like those numbers or iconic pictures – support the story making process that takes place when a public discourse around scientific data emerges. Public debates about events or topics are non-linear processes, interpretation of the known speculative facts is itself subject to discussion, and it is a struggle over meaning. Interpretation is made within different arenas and with differing accentuations.

Lessons Learned from the State of Research

For the quest of putting forward an analytical scheme for cultural narrative analysis there are some lessons to be learned from this overview of studies in the social sciences take on climate change:

(1) Understanding public perception of climate change needs to consider the cultural fabric where it takes place. That is that there is merit to the idea that culture is not to be seen as an add-on to social science research in this area. Rather, a cultural perspective can allow for an understanding the patterns that lie beneath social responses to climate change and its challenges. Climate change research might therefore benefit from a sociological approach, that understands the topic at hand as deeply cultural in nature.
(2) Social sciences are aware of the importance of narratives, especially for complicated phenomena like climate change. However, research here seems to be somewhat all over the place with no clear analytical framework. This is evident foremost in the varying use of terminology (narrative, story, frame, etc.), but also in the lack of a grounded theoretical approach. Going forward with inquiries into public perception of climate change and other complex social issues, a theoretical concept for analyzing social narratives might allow for comparisons over time and national contexts as well as for systemizing research interests.
(3) Research that investigates narratives involving climate change, heavily focuses on the reception-side of a message. That is, research is mainly guided by the quest to understand how a specific audience reacts to a story, how people can be persuaded or how the media fills its role as mediator between messenger and audience. While this is a very important and vital part of our understanding of the global crisis and while it considers the relevance of communication, it omits the role of the messenger. It seems that the place where stories about the fight against climate change originate, is somewhat of a blind spot in communication and social sciences research.

Considering all three aspects, the following contribution to the field of climate change communication views the topic of climate change narratives as cultural per se; to see culture not as a variable and material object of study, but as a way to understand how social meaning takes place. And if we take into account that, without social meaning, climate change is just a string of numbers and models, a cultural approach to the topic seems to be a viable path forward. The concept of narratives provides a tool for this endeavor as it allows for understanding social action and social meaning as cultural text (Geertz 1973; Smith 2011).

Notes

1. Boholm differentiates three basic modes of knowledge about risk: everyday experience, science driven scenarios, and collective narratives.
2. For a profound overview of studies that have explored media attention to climate change, see Schmidt et al. (2013: 1235–1237).
3. IPCC is the abbreviation for Intergovernmental Panel on Climate Change, an intergovernmental body under the auspices of the United Nations set up to assess and evaluate scientific findings in the field of climate change.
4. In 1988, climatologist James E. Hansen elaborated the trend of global warming before Congressional committees in the USA. His testimony helped to raise broad awareness of the issue.
5. Alternatives to Downs: Ungar's explanation of attention to global climate change focuses on the social scare that the hot summer of 1988 precipitated: real world events attract social attention, the real world impacts of the drought of 1988 brought global warming into view as a legitimate threat to personal wellbeing. This social scare catalyzes demand for news, to which the media responded (Ungar 2007).
6. Concerning journalistic influence on scientific reporting also refer to the realm of gatekeeper research, e.g. Robinson (1973).
7. For an in-depth analysis of the lobbying efforts against this anti-smoking-development see Oreskes and Conway (2011).
8. The veil of ignorance implies that a just society would be possible if no member knows anything about the background or motifs of the other members (such as race, gender, social background). This way all members of a society would have to agree on a set of rules that are not compromised by special interests (Rawls 1999).
9. This study stands out among most of other media studies since it does not examine newspaper articles or TV reports, but looks into the impact of popular culture, i.e. movies, and shall thus be described here to more detail: Lowe et al. interviewed focus groups in the UK before and after a viewing of the blockbuster movie "The day after tomorrow", which depicts consequences of an abrupt climatic change, giving rise to a new ice age. The researchers asked respondent to estimate the likelihood of extreme impacts, their overall concern about climate change, their motivation to personally take action and the just distribution of responsibility for the problem of

climate change. Even though most viewers saw the movie as fiction and not scientific fact, a significant amount was more committed to taking action than before seeing the movie. But, this movie as well works with fear and the authors here come to a similar conclusion as Aronson (2008), Ereaut and Segnit (2006), and others, which is that a terrifying message is not helping to change behavior if precise, effective, and doable strategies are offered.

10. In the debate about climate change the term climate justice argues for a proportional burden sharing of the costs the ecological crisis will cause (Birnbacher 2010).
11. This modern understanding of nature, rooted in the beginnings of the industrial revolution, supersedes a cultural pattern of men that originally did not put nature into an inferior role. Hunger (Hunger and Wilkens 2010) claims that climate change will force societies to re-think their relationship with nature and the value of its resources. This deduction might be depicted as naïve, since work on technical solutions to global warming is already undertaken. It also takes off from the assumption of a balanced nature-man-relation and the industrial revolution as a turning point within this relation, thereby omitting Christian heritage of 'govern the earth' that influenced western societies' culture and handling of nature.
12. See also Hoffman (2011) for an explicit treatment of climate skeptics.
13. Here, the fatalist's worldview is missing from the analysis, the authors do not offer an explanation for this. Jones, at another occasion, simply remarks that "it is common in CT scholarship to exclude fatalists form analysis" (Jones 2010: 68). A possible, however weak, explanation at least in this context could be that the fatalist's worldview is more introversive and does not influence social life as much as the other's do.
14. On a more specific aspect of social consequences of global warming, Bettini (2012) examines storylines that emerge in the discourse about climate refugees. Based on an analysis of different reports, the study detects four discursive families: a capitalist, a humanitarian, a radical, and a scientific discourse. Even though the study is purely explorative – dealing with only four reports – it is worth mentioning in this context, since it is looking into discursive dealing with climate change.
15. Heyd focuses not only on the perception side but even further on the cultural possibilities of answers to climate change.
16. Additional work on media usage and construction of narratives can be found in Jacobs (1996), Gronbeck (1983), and Darnton (1974).
17. As a result of the 2010 UN climate summit in Cancún the UN member states agreed on the target that the average global surface temperature must not increase of 2 °C over the pre-industrial average to avoid dangerous anthropogenic interference with the climate system.
18. UNFCCC is the abbreviation for United Nations Framework Convention on Climate Change.

REFERENCES

Alexander, J. C. (Ed.). (2003). *The Meanings of Social Life. A Cultural Sociology.* Oxford/New York: Oxford University Press. Retrieved from http://www.worldcat.org/oclc/51258793

Aronson, E. (2008). Fear, Denial, and Sensible Action in the Face of Disasters. *Social Research, 75*(3), 855–872.

Bauer, M. W., Allum, N., & Miller, S. (2007). What Can We Learn from 25 Years of PUS Survey Research? Liberating and Expanding the Agenda. *Public Understanding of Science, 16*(1), 79–95. https://doi.org/10.1177/0963662506071287.

Becker, E., & Jahn, T. (2006). *Soziale Ökologie. Grundzüge einer Wissenschaft von den gesellschaftlichen Naturverhältnissen.* Frankfurt am Main/New York: Campus.

Bell, A. (1991). *The Language of News Media.* Oxford/Cambridge, MA: Blackwell.

Bell, A. (1994). Climate of Opinion: Public and Media Discourse on the Global Environment. *Discourse & Society, 5*(1), 33–64. https://doi.org/10.1177/0957926594005001003.

Bettini, G. (2012). Climate Barbarians at the Gate? A Critique of Apocalyptic Narratives on "Climate Refugees". *Geoforum, 45,* 63–72.

Bickerstaff, K., & Walker, G. (1999). Clearing the Smog: Public Response to Air-Quality Information. *Local Environment, 4*(3), 279–294.

Birnbacher, D. (2010). Klimaverantwortung als Verteilungsproblem. In H. Welzer, H.-G. Soeffner, & D. Giesecke (Eds.), *KlimaKulturen. Soziale Wirklichkeiten im Klimawandel* (pp. 111–127). Frankfurt am Main: Campus.

Boholm, Å. (2015). *Anthropology and Risk. Earthscan Risk in Society Series.* London: Routledge/Taylor & Francis Group.

Bostrom, A., Morgan, M. G., Fischhoff, B., & Read, D. (1994). What Do People Know About Global Climate Change? 1. Mental Models. *Risk Analysis, 14*(6), 959–970. https://doi.org/10.1111/j.1539-6924.1994.tb00065.x.

Boykoff, M. T. (2008). Lost in Translation? The United States Television News Coverage of Anthropogenic Climate Change, 1995–2004. *Climatic Change, 86,* 1–11.

Boykoff, M. T., & Boykoff, J. M. (2004). Balance as Bias. Global Warming in the US Prestige Press. *Global Environmental Change, 14*(2), 125–136.

Boykoff, M. T., & Boykoff, J. M. (2007). Climate Change and Journalistic Norms: A Case-Study of US Mass-Media Coverage. *Geoforum, 38*(6), 1190–1204. https://doi.org/10.1016/j.geoforum.2007.01.008.

Bravo, M. T. (2009). Voices from the Sea Ice: The Reception of Climate Impact Narratives. *Journal of Historical Geography, 35,* 256–278.

Brossard, D., Shanahan, J., & McComas, K. (2004). Are Issue-Cycles Culturally Constructed? A Comparison of French and American Coverage of Global Warming. *Mass Communication and Society, 7*(3), 359–377.

Bundeszentrale für gesundheitliche Aufklärung (BZgA). (2016). Die Drogenaffinität Jugendlicher in der Bundesrepublik Deutschlanf 2015: Rauchen, Alkoholkonsum und Konsum illegaler Drogen: aktuelle Verbreitung und Trends. Berlin.
Büscher, C. (2010). Formen ökologischer Aufklärung. In C. Büscher (Ed.), *Ökologische Aufklärung. 25 Jahre "Ökologische Kommunikation"* (pp. 19–49). Wiesbaden: VS Verlag für Sozialwissenschaften.
Büscher, C., & Japp, K. P. (2010). Vorwort. In C. Büscher (Ed.), *Ökologische Aufklärung. 25 Jahre "Ökologische Kommunikation"* (pp. 19–49). Wiesbaden: VS Verlag für Sozialwissenschaften.
Carvalho, A., & Burgess, J. (2005). Cultural Circuits of Climate Change in U.K. Broadsheet Newspapers, 1985–2003. *Risk Analysis, 25*(6), 1457–1469.
Chai, S.-K., Liu, M., & Kim, M.-S. (2009). Cultural Comparisons of Beliefs and Values: Applying the Grid-Group Approach to the World Values Survey. *Beliefs and Values, 1*(2), 193–208. https://doi.org/10.1891/1942-0617.1.2.193.
Chakrabarty, D. (2010). Das Klima der Geschichte: Vier Thesen. In H. Welzer, H.-G. Soeffner, & D. Giesecke (Eds.), *KlimaKulturen. Soziale Wirklichkeiten im Klimawandel* (pp. 270–301). Frankfurt am Main: Campus.
Clausen, L. (2010). Wohin mit den Klimakatastrophen. In H. Welzer, H.-G. Soeffner, & D. Giesecke (Eds.), *KlimaKulturen. Soziale Wirklichkeiten im Klimawandel* (pp. 97–110). Frankfurt am Main: Campus.
Council of Canadian Academies. (2014). Report from the Expert Panel on Evaluating the Effectiveness of Health Product Risk Communication. Preliminary Version.
Dahl, T. (2015). Contested Science in the Media: Linguistic Traces of News Writers' Framing Activity. *Written Communication, 32*(1), 39–65. https://doi.org/10.1177/0741088314557623.
Daniels, S., & Endfield, G. H. (2009). Narratives of Climate Change: Introduction. *Journal of Historical Geography, 35*(2), 215–222.
Darnton, R. (1974). Writing News and Telling Stories. *Daedalus, 104*(2), 175–194. Retrieved from http://www.jstor.org/stable/20024337
Daschkeit, A., & Dombrowsky, W. R. (2010). Die Realität einer Katastrophe. Gesellschaftliche Diskurse zum Klimawandel. In C. Büscher (Ed.), *Ökologische Aufklärung. 25 Jahre Ökologische Kommunikation* (pp. 69–95). Wiesbaden: VS Verlag für Sozialwissenschaften.
Douglas, M. (1982). *In the Active Voice*. London/Boston: Routledge & K. Paul.
Douglas, M. (2004). *Purity and Danger: An Analysis of the Concepts of Pollution and Taboo*. London: Routledge. (Original Work Published 1966).
Douglas, M., & Wildavsky, A. B. (1983, c1982). *Risk and Culture: An Essay on the Selection of Technical and Environmental Dangers* (1st ed.). Berkeley: University of California Press.

Downing, P., & Ballantyne, J. (2007). *Tipping Point or Turning Point? Social Marketing and Climate Change*. London: IPSOS Mori Social Research Institute.

Downs, A. (1972). Up and Down with Ecology – The "Issue-Attention-Cycle". *The Public Interest, 28*, 38–51.

Dunlap, R. E., & McCright, A. M. (2008). A Widening Gap: Republican and Democrativ Views on Climate Change. *Environment: Science and Policy for Sustainable Development, 50*(5), 26–35.

Ereaut, G., & Segnit, N. (2006). *Warm Words: How Are We Telling the Climate Story and Can We Tell It Better?* London. Retrieved from Institute for Public Policy Research website: http://www.ippr.org/images/media/files/publication/2011/05/warm_words_1529.pdf

Espeland, W., & Micheals, S. (2012, January). The History of 10%: Social Science Measures and the Construction of Gay Identities. Yale University, C. for Cultural Sociology. CCS Workshop, New Haven.

Eyerman, R., & Jamison, A. (1991). *Social Movements. A Cognitive Approach*. Cambridge: Polity Press.

Fischer-Kowalski, M. (1997). Society's Metabolism: On the Childhood and Adolescence of a Rising Conceptual Star. In M. R. Redclift & G. Woodgate (Eds.), *The International Handbook of Environmental Sociology* (pp. 119–137). Cheltenham/Northampton: Edward Elgar.

Fischer-Kowalski, M., & Weisz, H. (2005). Society as Hybrid Between Material and Symbolic Realms. Toward a Theoretical Framework of Society-Nature Interaction. In M. R. Redclift & G. Woodgate (Eds.), *New Developments in Environmental Sociology* (pp. 113–146). Cheltenham: Elgar.

Fischhoff, B. (1995). Risk Perception and Communication Unplugged: Twenty Years of Process. *Risk Analysis, 15*(2), 137–145.

Galtung, J., & Ruge Mari, H. (1965). The Structure of Foreign News. *Journal of International Peace Research, 2*, 64–90.

Gans, H. J. (2005). *Deciding What's News: A Study of CBS Evening News, NBC Nightly News, Newsweek, and Time*. Evanston: Northwestern University Press.

Gardner, G. T., & Stern, P. C. (2002). *Environmental Problems and Human Behavior* (2nd ed.). Boston: Pearson Custom Publishing.

Geertz, C. (1973). *The Interpretation of Cultures. Selected Essays*. New York: Basic Books.

Giddens, A. (2009). *The Politics of Climate Change, Cambridge*. Malden: Polity Press.

Goidel, R. K., Shields, T. G., & Peffley, M. (1997). Priming Theory and RAS Models: Toward an Integrated Perspective of Media Influence. *American Politics Quarterly, 25*, 287–318.

Gronbeck, B. E. (1983). II. Narrative, Enactment, and Television Programming. *Southern Speech Communication Journal, 48*(3), 229–243. https://doi.org/10.1080/10417948309372567.

Grundmann, R. (2007). Climate Change and Knowledge Politics. *Environmental Politics, 16*(3), 416–434.
Grundmann, R., & Krishnamurthy, R. (2010). The Discourse of Climate Change: A Corpus-based Approach. *Critical Approaches to Discourse Analysis Across Disciplines, 4*(2), 125–146.
Hagner, M. (2010). Haben die Geisteswissenschaften die Zukunft vergessen? In H. Welzer, H.-G. Soeffner, & D. Giesecke (Eds.), *KlimaKulturen. Soziale Wirklichkeiten im Klimawandel* (pp. 20–32). Frankfurt am Main: Campus.
Hamblyn, R. (2009). The Whistleblower and the Canary: Rhetorical Construction of Climate Change. *Journal of Historical Geography, 35,* 223–236.
Hart, P. S., & Nisbet, E. C. (2011). Boomerang Effects in Science Communication: How Motivated Reasoning and Identity Cues Amplify Opinion Polarization About Climate Mitigation Policies. *Communication Research.* Advance Online Publication. https://doi.org/10.1177/0093650211416646
Heidbrink, L. (2010). Kultureller Wandel: Zur kulturellen Bewältigung des Klimawandels. In H. Welzer, H.-G. Soeffner, & D. Giesecke (Eds.), *KlimaKulturen. Soziale Wirklichkeiten im Klimawandel* (pp. 49–64). Frankfurt am Main: Campus.
Heyd, T. (2010). Climate Change, Individual Responsibilities and Cultural Frameworks. *Human Ecology Review, 17*(2), 86–95.
Hinchliffe, S. (1996). Helping the Earth Begins at Home. The Social Construction of Socio-Environmental Responsibilities. *Global Environmental Change, 6*(1), 53–62.
Hoffman, A. J. (2010). Climate Change as a Cultural and Behavioral Issue: Addressing Barriers and Implementing Solutions. *Organizational Dynamics, 39*(4), 295–305.
Hoffman, A. J. (2011). The Culture and Discourse of Climate Skepticism. *Strategic Organization, 9*(1), 77–84.
Hunger, B., & Werner, W. (2010). Architektur und Städtebau im Spannungsfeld von klimakultureller Prägung und sozialökonomischer Entwicklung. In H. Welzer, H.-G. Soeffner, & D. Giesecke (Eds.), *KlimaKulturen. Soziale Wirklichkeiten im Klimawandel* (pp. 161–179). Frankfurt am Main: Campus.
Jacobs, R. N. (1996). Producing the News, Producing the Crisis: Narrativity, Television and News Work. *Media, Culture & Society, 18*(3), 373–397. https://doi.org/10.1177/016344396018003002.
Joch Robinson, G. (1973). 25 Jahre "Gatekeeper" Forschung. Eine kritische Rückschau und Bewertung. In J. Aufermann, H. Bohrmann, & R. Sülzer (Eds.), *Gesellschaftliche Kommunikation und Information. Forschungsrichtungen und Problemstellungen; ein Arbeitsbuch zur Massenkommunikation* (pp. 344–355). Frankfurt a.M: Athenäum Fischer Taschenbuch Verl.
Jones, M. D. (2010). *Heroes and Villains: Cultural Narratives, Mass Opinions, and Climate Change* (Unpublished Dissertation). University of Oklahoma, Norman, Oklahoma.

Jones, M. D. (2011). Leading the Way to Compromise? Cultural Theory and Climate Change Opinion. *PS: Political Science and Politics, 44*(4), 720–725.

Keller, C., Siegrist, M., & Gutscher, H. (2006). The Role of the Affect and Availability Heuristics in Risk Communication. *Risk Analysis, 26*(3), 631–639.

Kellstedt, P. M., Zahran, S., & Vedlitz, A. (2008). Personal Efficacy, the Information Environment, and Attitudes Toward Global Warming and Climate Change in the United States. *Risk Analysis, 28*(1), 113–126. https://doi.org/10.1111/j.1539-6924.2008.01010.x.

Kollmuss, A., & Agyeman, J. (2002). Mind the Gap: Why Do People Act Environmentally and What Are the Barriers to Pro-environmental Behavior? *Environmental Education Research, 8*(3), 239–260.

Krauss, W., & Storch, H. von. (2005). Culture Contributes to Perception of Climate Change. *Nieman Reports, 59*(4). Retrieved from http://www.nieman.harvard.edu/reports/article/100600/Culture-Contributes-to-Perceptions-of-Climate-Change.aspx

Krimsky, S. (2007). Risk Communication in the Internet Age: The Rise of Disorganized Skepticism. *Environmental Hazards, 7*(2), 157–164.

Kuckartz, U. (2010). Nicht hier, nicht jetzt, nicht ich – Über die symbolische Bearbeitung eines ernsten Problems. In H. Welzer, H.-G. Soeffner, & D. Giesecke (Eds.), *KlimaKulturen. Soziale Wirklichkeiten im Klimawandel* (pp. 144–160). Frankfurt am Main: Campus.

Leggewie, C., & Welzer, H. (2010). *Das Ende der Welt, wie wir sie kannten: Klima, Zukunft und die Chancen der Demokratie (4. Aufl.).* Frankfurt am Main: Fischer.

Leiserowitz, A. A. (2003). *Global Warming in the American Mind: The Roles of Affect, Imagery, and Worldviews in Risk Perception, Policy Preferences, and Behavior* (Unpublished Dissertation). University of Oregon, Eugene. Retrieved from http://decisionresearch.org/pdf/540.pdf

Leiserowitz, A. A. (2005). American Risk Perceptions: Is Climate Change Dangerous? *Risk Analysis, 25*(6), 1433–1442.

Leiserowitz, A. A. (2007). Communicating the Risks of Global Warming: American Risk Perceptions, Affective Images, and Interpretive Communities. In S. C. Moser & L. Dilling (Eds.), *Creating a Climate for Change: Communicating Climate Change and Facilitating Social Change* (pp. 44–63). New York: Cambridge University Press.

Leiserowitz, A. A., & Smith, N. (2010). *Knowledge of Climate Change Across Global Warming's Six Americas.* New Haven: Yale University.

Leiserowitz, A. A., Maibach, E. W., Roser-Renouf, C., & Smith, N. (2010). *Climate Change in the American Mind: Americans' Global Warming Beliefs and Attitudes in June 2010.* Yale University and George Mason University. New Haven, CT: Yale Project on Climate Change Communication.

Leiss, W. (1996). Three Phases in the Evolution of Risk Communication Practice. *The ANNALS of the American Academy of Political and Social Science, 545*(1), 85–94.
Lever-Tracy, C. (2008). Global Warming and Sociology. *Current Sociology, 56*(3), 445–466. https://doi.org/10.1177/0011392107088238.
Liverman, D. M. (2009). Conventions of Climate Change: Constructions of Danger and the Dispossession of the Atmosphere. *Journal of Historical Geography, 35*(2), 279–296.
Lowe, T., Brown, K., Dessai, S., de Franca, M., Haynes, K., & Vincent, K. (2006). Does Tomorrow Ever Come? Disaster Narrative and Public Perceptions of Climate Change. *Public Understanding of Science, 15*(4), 435–457. https://doi.org/10.1177/0963662506063796.
Maibach, E. W., Nisbet, M. C., Baldwin, P., Akerlof, K., & Diao, G. (2010). Reframing Climate Change as a Public Health Issue: An Exploratory Study of Public Reactions. *BMC Public Health, 10*(1), 299. https://doi.org/10.1186/1471-2458-10-299.
Marshall, G. (2014). *Hearth and Hierath: Constructing Climate Change Narratives Around National Identity*. Climate Outreach and Information Network, Oxford.
Mazur, A. (1998). Global Environmental Change in the News – 1987–1990 vs. 1992–1996. *International Sociology, 13*(4), 457–472.
McComas, K., & Shanahan, J. (1999). Telling Stories About Climate Change: Measuring the Impact of Narratives on Issue Cycles. *Communication Research, 26*(30), 30–57.
Messner, D. (2010). Globale Strukturanpassung: Weltwirtschaft und Weltpolitik in den Grenzen des Erdsystems. In H. Welzer, H.-G. Soeffner, & D. Giesecke (Eds.), *KlimaKulturen. Soziale Wirklichkeiten im Klimawandel* (pp. 65–80). Frankfurt am Main: Campus.
Michaelis, L. (2000). European Narratives About Human Nature, Society, and the Good Life. In E. Jochem, J. A. Sathaye, & D. Bouille (Eds.), *Society, Behaviour, and Climate Change Mitigation* (pp. 157–168). New York [etc.]: Kluwer.
Minkmar, N. (2010). Der Pfirsich in Paris – Ein Essay über die Klimakultur des französischen Südwestens. In H. Welzer, H.-G. Soeffner, & D. Giesecke (Eds.), *KlimaKulturen. Soziale Wirklichkeiten im Klimawandel* (pp. 212–221). Frankfurt am Main: Campus.
Moser, S. C., & Dilling, L. (2004). Making Climate Hot: Communicating the Urgency and Challenge of Global Climate Change. *Environment: Science and Policy for Sustainable Development, 46*(10), 32–46.
Moser, S. C., & Dilling, L. (Eds.). (2007). *Creating a Climate for Change: Communicating Climate Change and Facilitating Social Change*. New York: Cambridge University Press.

Moser, S. C., & Dilling, L. (2011). Communicating Climate Change: Closing the Science-Action Gap. In J. S. Dryzek, R. B. Norgaard, & D. Schlosberg (Eds.), *Oxford Handbook of Climate Change and Society* (pp. 161–174). Oxford,/New York: Oxford University Press.

The Canada Institute of the Woodrow, Moser, S. C., & Walser, M. (2008). Communicating Climate Change Motivating Citizen Action. In: *Encyclopedia of Earth*. Cutler J. Cleveland (Ed.) Washington, D.C.: Environmental Information Coalition, National Council for Science and the Environment.

Myers, T. A., Nisbet, M. C., Maibach, E. W., & Leiserowitz, A. A. (2012). A Public Health Frame Arouses Hopeful Emotions About Climate Change. A Letter. *Climatic Change, 113*(3–4), 1105–1112. https://doi.org/10.1007/s10584-012-0513-6.

Nelkin, D. (1987). *Selling Science: How the Press Covers Science and Technology.* New York: W.H. Freeman.

Ney, S., & Thompson, M. (2000). Cultural Discourses in the Global Climate Change Debate. In E. Jochem, J. A. Sathaye, & D. Bouille (Eds.), *Society, Behaviour, and Climate Change Mitigation* (pp. 65–92). New York [etc.]: Kluwer. Retrieved from http://link.springer.com/book/10.1007/0-306-48160-X/page/1

Nicholson-Cole, S. A. (2005). Representing Climate Change Futures: A Critique on the Use of Images for Visual Communication. *Computers, Environment and Urban Systems, 29*(3), 255–273.

Nisbet, M. C. (2005). The Competition of Worldviews: Values, Information, and Public Support for Stem Cell Research. *International Journal of Public Opinion Research, 17*(1), 90–112.

Nisbet, M. C. (2009). Communicating Climate Change. Why Frames Matter for Public Engagement. *Environment: Science and Policy for Sustainable Development, 51*(2), 12–23.

Nisbet, M. C., & Goidel, R. K. (2007). Understanding Citizen Perception of Science Controversy: Bridging the Ethnographic-Survey Research Divide. *Public Understanding of Science, 16*, 421–440.

Nissani, M. (1999). Media Coverage of the Greenhouse Effect. *Population and Environment, 21*(1), 27–43. https://doi.org/10.1007/BF02436119.

Norgaard, K. M. (2011). *Living in Denial: Climate Change, Emotions, and Everyday Life.* Cambridge, MA: MIT Press.

O'Neill, S., & Hulme, M. (2009). An Iconic Approach for Representing Climate Change. *Global Environmental Change, 19*(4), 402–410.

O'Neill, S., & Nicholson-Cole, S. A. (2009). "Fear Won't Do It": Promoting Positive Engagement With Climate Change Through Visual and Iconic Representations. *Science Communication, 30*(3), 355–379.

Oreskes, N., & Conway, E. M. (2011). *Merchants of Doubt: How a Handful of Scientists Obscured the Truth on Issues from Tobacco Smoke to Global Warming.* New York: Bloomsbury Press.

Osbaldiston, N. (2010). What Role Can Cultural Sociology Play in Climate Change Adaptation? In Cultural Sociology Group (Ed.), *Cultural Fields. Newsletter of the TASA Cultural Sociology Thematic Group*, No. 2, pp. 6–7.
Ostrom, E. (1990). *Governing the Commons. The Evolution of Institutions for Collective Action*. Cambridge/New York: Cambridge University Press.
Owens, S. (2000). 'Engaging the Public': Information and Deliberation in Environmental Policy. *Environment and Planning, 32*(7), 1141–1148. https://doi.org/10.1068/a3330.
Owens, S., & Driffill, L. (2008). How to Change Attitudes and Behaviors in the Context of Energy. *Energy Policy, 36*(12), 4412–4418.
Paschen, J.-A., & Ison, R. (2014). Narrative Research in Climate Change Adaptation – Exploring a Complementary Paradigm for Research and Governance. *Research Policy, 43*(6), 1083–1092.
Pearce, W., Brown, B., Nerlich, B., & Koteyko, N. (2015). Communicating Climate Change: Conduits, Content, and Consensus. *Wiley Interdisciplinary Reviews: Climate Change, 6*(6), 613–626. https://doi.org/10.1002/wcc.366.
Petersen, K. K. (2009). Revisiting Downs' Issue-Attention Cycle: Revisiting Downs' Issue Attention Cycle: International Terrorism and U.S. Public Opinion. *Journal of Strategic Security, 2*(4), 1–16.
Post, S. (2008). *Klimakatastrophe oder Katastrophenklima? Die Berichterstattung über den Klimawandel aus Sicht der Klimaforscher*. Frankfurt am Main: Fischer.
Priddat, B. P. (2010). Klimawandel: Das Ende der geotopologischen Identität. In H. Welzer, H.-G. Soeffner, & D. Giesecke (Eds.), *KlimaKulturen. Soziale Wirklichkeiten im Klimawandel* (pp. 81–96). Frankfurt am Main: Campus.
Rahmstorf, S., & Schellnhuber, H. J. (2007). Der Klimawandel: Diagnose, Prognose, Therapie (6. Aufl., Orig.-Ausg.). Beck'sche Reihe C.-H.-Beck-Wissen: Vol. 2366. München: Beck. Retrieved from http://www.gbv.de/dms/faz-rez/S13200706171109587.pdf
Rawls, J. (1999). *A Theory of Justice* (Rev. ed.). Cambridge, MA: Belknap Press of Harvard University Press.
Read, D., Bostrom, A., Morgan, M. G., Fischhoff, B., & Smuts, T. (1994). What Do People Know About Climate Change? 2. Survey Studies of Educated Laypeople. *Risk Analysis, 14*(6), 971–982.
Renn, O. (2008). *Risk Governance. Coping with Uncertainty in a Complex World*. London/Sterling: Earthscan.
Renn, O. (2011). The Social Amplification/Attenuation of Risk Framework: Application to Climate Change. *Wiley Interdisciplinary Reviews: Climate Change, 2*(2), 154–169.
Schäfer, M., Ivanova, A., & Schmidt, A. (2011). Global Climate Change – Global Public Sphere? Media Attention for Climate Change in 23 Countries. *Studies in Communication/Media, 0*(1), 131–148. http://www.scm.nomos.de/fileadmin/scm/doc/SCM_11_01_05.pdf

Schmidt, A., Ivanova, A., & Schäfer, M. S. (2013). Media Attention for Climate Change Around the World: A Comparative Analysis of Newspaper Coverage in 27 Countries. *Global Environmental Change, 23*(5), 1233–1248. https://doi.org/10.1016/j.gloenvcha.2013.07.020.

Schwarz, M., & Thompson, M. (1990). *Divided We Stand.: Redefining Politics, Technology, and Social Choice.* Philadelphia: University of Pennsylvania Press.

Semenza, J. C., Hall, D. E., Wilson, D. J., Bontempo, B. D., Sailor, D. J., & George, L. A. (2008). Public Perception of Climate Change. *American Journal of Preventive Medicine, 35*(5), 479–487. https://doi.org/10.1016/j.amepre.2008.08.020.

Shanahan, J., McComas, K., & Pelstring, L. (1999). Using Narratives to Think About Environmental Attitude and Behavior: An Exploratory Study. *Society & Natural Resources, 12*(5), 405–419. https://doi.org/10.1080/089419299279506.

Smith, P. (2001). *Cultural Theory: An Introduction.* Malden/Oxford: Blackwell. Retrieved from http://www.worldcat.org/oclc/491877587

Smith, P. (2005). *Why War? The Cultural Logic of Iraq, the Gulf War, and Suez.* Chicago: University of Chicago Press.

Smith, P. (2011). The Balinese Cockfight Decoded: Reflections on Geertz and Structuralism. In C. A. Jeffrey et al. (Eds.), *Interpreting Clifford Geertz. Cultural Investigation in the Social Sciences* (pp. 17–32). New York: Palgrave Macmillan.

Smith, P. (2012). Narrating Global Warming. In J. C. Alexander, R. N. Jacobs, & P. Smith (Eds.), *The Oxford Handbook of Cultural Sociology* (pp. 745–760). New York: Oxford University Press.

Smith, P., & West, B. (1996). Drought, Discourse, and Durkheim: A Research Note. *Australian and New Zealand Journal of Sociology, 32*(1), 93–102.

Smith, P., & West, B. (1997). Natural Disasters and National Identity: Time, Space, and Mythology. *Australian and New Zealand Journal of Sociology, 33*(2), 205–215.

Takahashi, B. (2009). Social Marketing for the Environment: An Assessment of Theory and Practice. *Applied Environmental Education & Communication, 8*(2), 135–145. https://doi.org/10.1080/15330150903135889.

Thompson, M. (2003). Cultural Theory, Climate Change, and Clumsiness. *Economic and Political Weekly, 38*(48), 5107–5113.

Thompson, M., Ellis, R., & Wildavsky, A. (1990). *Cultural Theory.* Boulder [etc.]: Westview Press.

Trumbo, C. W. (1996). Constructing Climate Change: Claims and Frames in US News Coverage of an Environmental Issue. *Public Understanding of Science, 5*(3), 269–283.

Trumbo, C. W., & Shanahan, J. (2000). Social Research on Climate Change: Where We Have Been, Where We Are, and Where We Might Go. *Public Understanding of Science, 9*(3), 199–204.

Tuchman, G. (1978). *Making News: A Study in the Construction of Reality.* New York: The Free Press.
Ungar, S. (2007). Public Scares: Changing the Issue Culture. In S. C. Moser & L. Dilling (Eds.), *Creating a Climate for Change: Communicating Climate Change and Facilitating Social Change* (pp. 81–88). New York: Cambridge University Press.
Verweij, M., Douglas, M., Ellis, R., Engel, C., Hendriks, F., Lohmann, S., Thompson, M. (2006). Clumsy Solutions for a Complex World: The Case of Climate Change. *Public Administration, 84*(4), 817–843. https://doi.org/10.1111/j.1540-8159.2005.09566.x-i1
Weingart, P., Engels, A., & Pansegrau, P. (2000). Risks of Communication: Discourses on Climate Change in Science, Politics, and the Mass Media. *Public Understanding of Science, 9*(3), 261–283. https://doi.org/10.1088/0963-6625/9/3/304.
Welzer, H., Soeffner, H.-G., & Giesecke, D. (Eds.). (2010). *KlimaKulturen: Soziale Wirklichkeiten im Klimawandel.* Frankfurt am Main: Campus. Retrieved from http://deposit.d-nb.de/cgi-bin/dokserv?id=3406997&prov=M&dok_var=1&dok_ext=htm
Whitmarsh, L. (2008). Are Flood Victims More Concerned About Climate Change Than Other People? The Role of Direct Experience in Risk Perception and Behavioral Response. *Journal of Risk Research, 11*(3), 351–374.
Wiesenthal, H. (2010). Klimawandel der Umweltpolitik? Oder: Energiekonzepte als Identitätskrücke. In C. Büscher (Ed.), *Ökologische Aufklärung. 25 Jahre "Ökologische Kommunikation"* (pp. 173–202). Wiesbaden: VS Verlag für Sozialwissenschaften.
Wilson, K. M. (1995). Mass Media and Sources of Global Warming Knowledge. *Mass Communications Review, 22*(1; 2), 75–89.

CHAPTER 3

How to Understand the Role of Narratives in Environmental Communication: Cultural Narrative Analysis

Abstract This chapter discusses different approaches in the social sciences dealing with narratives and introduces the reader to the narrative terminology. It will debate the social function of narratives and focus especially on narrative research in the realm of environmental problems to provide a basis for the ensuing analysis and findings. Subsequently, this chapter elaborates on the theoretical foundations that underlie the narrative analysis, drawing on cultural sociology and narrative theory. It suggests an analytical scheme to investigate cultural narratives with the help of narrative and literary theory.

Keywords Narrative analysis • Narrative theory • Genre • Narrative policy framework • Cultural narratives • Storytelling • Narratives turn

How we perceive the climate change is very much influenced by the cultural patterns it is embedded in. Narratives as an analytical device allow for investigating those basic cultural elements that structure our world views and our outlook on current and past events. However, narratives don't come natural to the discipline of social sciences or sociology, as the subject of interest centers on the fabric of societies, how people interact with each other and the environment, and how societies work. A cultural sociological perspective takes this empirical matter as a text to interpret collective meaning and its analysis can be seen analogous to that of a written text. Cultural sociology and environmental communication itself is thus well-advised to lean on insights form narrative theory and literary studies as guidelines through the exercise of narrative analysis. "Narrative provides a link between culture as system and culture as practice. […] Narrative is an arena in which meaning takes form, in which individuals connect to the public and social world" (Bonnell and Hunt 1999b: 17). Narrative analysis enables sociological research to avoid structural or cultural fundamentalism because it allows for understanding how culture and structure interact in shaping interests (Polletta 2006: 27).

Following a broad overview over narrative analysis within different disciplines, I will outline various attempts of defining narratives and with that describe structural and content elements of narratives. This provides the basis for a discussion of different models of narrative analysis: the Structural Model of Narrative by Labov and Waletzky (1997), Smith's Structural Model of Genre (2005), and Jones and McBeth's Narrative Policy Framework (2010). These models will be interrelated to cultural sociological demands of analysis. Starting from this, I will suggest an integrated attempt of cultural narrative analysis. This model will combine two ends of narrative analysis: formulaic and playful narrative research which are not to be understood as mutually exclusive (Smith 2007: 392–393).

Clearly narrative analysis is subject to various disciplines and in each of them are narratives analyzed in various forms. The term of narrative plays an important role in linguistics, literary theory, cultural theory, and social sciences. Since the early 1980s the use of narrative methods rose in the social sciences, from highly technical linguistic analysis of structure to interpretive approaches focusing on content (Hards 2012: 762; Elliott 2005: 3). Smith (Smith 2010: 132) points at two parallel outcomes of narrative analysis: the first refers to analytic and meta-narrative and thus depicts narratives as product of the analysis. Seeking answers to historical causalities, the researcher produces narratives in the course of the analysis. The second is described as ontological and public, cultural, or institutional narratives. Here, the narrator is in the focus, as she assigns meanings to events by sequencing them intentionally in a specific order. Elliot differentiates ontological narratives as first-order narratives from analytic narratives as second-order narratives. The latter are collective stories researcher's construct to make sense of the social world (Richardson 1990b: 25–6). Those collective stories however do not emerge from a cultural vacuum; rather, they are embedded in cultural structures. Sociological approach to narratives thus does not stay at the level of textual analysis: "it insists that story production and consumption is an empirical and social process involving stream of joint actions in local contexts themselves bound into wider negotiated social worlds. Texts are connected to lives, actions, contexts, and societies" (Plummer 1995: 24).

THE ROLE OF NARRATIVE IN SOCIAL ORGANIZATION

Franzosi (1998) poses the question how and why sociologists should be interested in narrative in the first place. After all, narratives seem to belong more in the realm of literary theorists than into the study of social action. On the other side, as the narrative turn indicates, narratives give way to uncover a collective story in sociologically significant terms (Richardson 1990a: 125). Franzosi states that "narrative texts are packed with sociological information, and a great deal of our empirical evidence is in narrative form" (Franzosi 1998: 517). Narrative is a useful concept for social research, since it is "not tied to any particular medium and it is independent of the distinction between fiction and non-fiction" (Ryan 2007: 26). Constructing a narrative plays an important role in social life, communities mobilize resources to protest, fight, or contribute to common projects around narratives (Smith 2010: 136). "Individual stories tie in with a society's narratives about collective cultural meanings. These in turn are embedded in what Lyotard called humankind's great 'meta-narratives' provided by religion, science and tradition" (Paschen and Ison 2014: 4). Real world events must be translated into narrative form. They "must be not only registered within chronological framework of their original occurrence but narrated as well, that is to say, revealed as possessing a structure, an order of meaning, that they do not possess as mere sequence" (White 1987: 5).

In attempts to a comprehensive definition, many authors both from literary studies as well as from the realm of social sciences and linguistics include the social role of narratives. Bonnell and Hunt describe the universality of narratives in daily social life: "narratives get their power from being woven into daily life – that is, by molding and expressing popular opinion of how individual motivation and action work" (Bonnell and Hunt 1999b: 18). Telling stories is a universal human activity (Hards 2012: 762), "human beings are storytelling animals, we tell stories about our triumphs and tragedies" (Alexander 2003: 84), human beings organize their experiences and memory in the form of narrative stories (Bruner 1991: 4), and these stories are everywhere in social life:

There are countless forms of narrative in the world. [...] Narrative is present in myth, legend, fables, tales, short stories, epics, history, tragedy, drame [suspense drama], comedy, pantomime, paintings (in Santa Ursula by Carpaccio for instance), stained-glass windows, movies, local news, conversation. Moreover, in this infinite variety of forms, it is present at all times, in all places, in all societies; indeed, narrative starts with the very history of mankind; there is not, there has never been anywhere, any people without narrative. [...]. Like life itself, it is there, international, transhistorical, transcultural. (Barthes 1975: 237)

However, Barthes should not be misunderstood in the sense that narratives simply exist in our world. Just like there is nothing natural about the way social actors behave among one another and how they perceive the world, likewise, "there is nothing natural about chronological registrations of events" (White 1987: 176). Thus, for the social researcher, narratives are perceived as result of human and social agency; as a way to make sense of the world and the events that occur in it. Thus, narratives are social meaning making in action (Paschen and Ison 2014: 4).

The following table gives an overview of quotes from different authors concerning the role of narratives in social life (Table 3.1):

The role of narratives contains methodical implications for their use in the social sciences. Hinchman and Hinchman state, that "narratives (stories) in the human sciences should be defined provisionally as discourses with a clear sequential order that connect events in a meaningful way for a definite audience and thus offer insights about the world and/ or people's experiences of it" (Hinchman and Hinchman 1997: xvi). Thus, narrative inquiry is a fruitful way for the social researcher to explore people's experiences and the meaning they connect to those experiences (Hards 2012: 762). Polletta sees the social use of singular narratives against the backdrop of familiar stories within a society and culture, thus influencing social structures. "The relationship between culture, structure, and story is thus complex and variable. Much of the time, structures are reproduced through stories that address familiar oppositions. Sometimes, stories undermine those oppositions in ways that mobilize overt change [...]. Stories of women having different job aspirations than men make sense, because they are heard against the backdrop of stories of women having different biologies than men and stories of little girls being different from little boys and stories of people having different tastes [...] (Polletta 2006: 15).

Interest in narrative analysis in the human and social sciences grew stronger since 1980s, where some scholars even see these disciplines taking

Table 3.1 The social role of narratives

Author(s)	Quote on the social role of narratives
Hards (2012: 762)	Narrative approaches suggest that people make sense of their experiences by telling stories to others and to themselves. Advocates claim that storytelling is a universal human activity.
Bonnell and Hunt (1999a: 17)	Narrative is an arena in which meaning takes form, in which individuals connect to the public and social world, and in which change therefore becomes possible.
Smith (2010: 129)	Actors articulate their beliefs and thoughts and conceive of appropriate actions to accompany those thoughts […] by the telling of a story with a beginning, middle, and end. Through these expressions, actors come to understand and construct their world and their place within it.
Elliott (2005: 3)	A narrative can be understood to organize a sequence of events into a whole so that the significance of each event can be understood through its relation to that whole. This way, a narrative conveys the meaning of events.
Dahlstrom (2010: 857)	Narratives influence what individuals believe about the world.
White (1987: 5)	Events must be […] narrated as well, that is to say, revealed as possessing a structure, an order of meaning that they do not possess as mere sequence.
Boholm (2015: 14–15)	Collective narratives about events […] are predominantly communicated through news media. […] there must be a story about intentions and motives, victims, villains, and heroes, all staged in a specific setting.
Bruner (1991: 4)	We organize our experience and our memory of human happenings mainly in the form of narrative […]. Unlike the constructions generated by logical and scientific procedures that can be weeded out by falsification, narrative constructions can only achieve verisimilitude. Narratives, then, are a version of reality whose acceptability is governed by convention and narrative necessity.

a "narrative turn" (Mishler 1995: 87–88; 117). Alexander identifies a growing interest in narrative analysis with sociologists "now reading literary theorists like Northrop Frye, Peter Brooks, and Frederic Jameson […]. The appeal of such theory lies partially in its affinity for a textual understanding of social life" (Alexander and Smith 2003b: 25). Narrative theory with a structuralist background works well within cultural sociology for it assures cultural autonomy in its analytic sense. A structuralist approach highlights the relationships between narrative elements (characters, plot, moral evaluation) in formal models, thus allowing for an application across cases without losing sight of each case's particularities (Alexander and Smith 2003a, b: 25–26).

Formal Definitions of Narrative

Definitions of narrative distinguish in general between structure and form on the one hand and content on the other. The basis for this distinction is a reference to Hayden White's content and form (1987), which can be seen analogous to Saussure's distinction between parole and langue and his semiotic use of signifier and signified, where the mere occurrence of events (specific dates, geographical details and so on) correspond as signified to the content of those events as the signifiers (White 1987: 9). Narrative scholars agree by and large on basic structural elements, which are:

- Beginning – middle – end
- Unfolding events
- Presentation of characters (hero – villain – victim)
- Plot
- Moral/transformation

Narratives consist of a beginning, a middle, and an end (Jones and McBeth 2010: 334; Smith 2010: 129; Richardson 2007: 146; Halttunen 1999: 165–166; Labov and Waletzky 1997: 12; White 1987: 17). "Everywhere people experience and interpret their lives in relationship to *time*. […] And, everywhere, humans make sense of their temporal worlds through the narrative" (Richardson 1990a: 124, italics original). Abbott introduces the term of narration complementary to narrative as the process of telling the story (Abbott 2007: 39–40). He thus achieves a distinction between story and plot as well as between story and narrative, where narrative becomes the representation of a story. This is important because it underlines the role narratives play in transporting meaning: a story can be plotted and narrated in different ways; the factual real-world events stay the same, but their meaning changes with the way they are narrated.

Ryan points out that there exists a general consensus on basic elements of a definition, such as a sequential order of presented events, causality between these events (as opposed to merely a list of events),[1] and those elements that render change possible within a narrative[2] (Ryan 2007: 23; Smith 2010: 133). Elliot summarizes those basic elements in her definition: "Narrative can be understood to organize a sequence of events into a whole so that the significance of each event can be understood through its relation to that whole" (Elliott 2005: 3). Stone states that these fundamental elements – temporal structure, characters, transformation – can also be found in the structure of policy problems (Stone 2002: 159).

The sequence of events is as important and influential to the transported meaning as the content of the narrative, because this "sequencing of events implies something about which events are necessary and contribute to a given outcome" (Smith 2010: 133). Besides the temporal paragraphs narratives contain a number of characters that play, as an individual or collective entity, a role in the events that are told (Margolin 2007: 66). The set of characters consists of one or more heroes as fixer(s) of a problem, one or more villains causing a problem, and one or more victims affected by the problem (McBeth et al. 2005: 415). Characters are part of the design that constitutes a plot and represents ideological positions (Richardson 2007). Missing from this list of characters is the role of the narrator or storyteller, which does not necessarily have to be one of the characters. In fact, as we will see in the analysis of climate change narratives, narrators position themselves differently within or outside the narrative they are telling. Analysis of structure helps to uncover this position. As the authority of the narrative the storyteller owns interpretational sovereignty (White 1987: 19). For this reason, the narrator should always be understood as part of the set of characters, irrespective of their actual position within the narrative (that is, if they play an active role in those events that are unfolding).

A Typology of Narrative Analysis

Broadly speaking, narrative analysis can focus on two parts of narratives: for one, on the process of segmentation, which identifies semiotic units, and secondly, on the process of integration, assembling those units of the process of segmentation into units of higher rank. The first process corresponds to the form, the second to the meaning (Barthes 1975: 266).

Mishler (1995) put forward a typology of narrative analysis. Narrative models thus focus on one of the three functions of language:

meaning – structure – interactional context.
These functions are analogous to the differentiation of
semantics – syntax – pragmatics and content – structure – performance.

The table below gives an overview of the interrelations of these three disciplines and their understanding of narratives (Table 3.2).

Table 3.2 Narrative in linguistics, literary theory, and social sciences

	Narrative			
Discipline Subject of analysis	Linguistic theory Structural elements of narratives	Literary theory Narratives in fictional texts	Social Sciences Narrative as individual life-stories	Analytic and meta-narratives

Source: Own illustration

With a combination of the models of narrative analysis that will be presented in the following chapters I will show the importance of all the above-mentioned levels of narrative analysis.

The following presentation of analytic models zooms out from the basic units of a narrative (Labov and Waletzky's structural model of narrative), through its external form and the meaning it transports through this form (Smith's structural model of genre), to the transmitted, literal content (Jones and McBeth's narrative policy framework), thus providing a line from structure to form to content alongside White's (1987) differentiation between content and form and Saussure's (1986) differentiation between parole and langue.

THE STRUCTURAL MODEL OF NARRATIVE: THE STRUCTURE OF NARRATIVE

The sociolinguists Waletzky and Labov undertake a formal and functional analysis of narrative (Labov and Waletzky 1997). Formal in the sense that it relies on the basic techniques of linguistic analysis, isolating structural units that correspond to a variety of superficial forms; functional in the sense that narratives are considered as one verbal technique for recapitulating experience (Labov and Waletzky 1997: 4). The model thus examines formal structural properties of narratives in relation to their social functions (Cortazzi 1993: 43). Understanding the structure of narrative "helps us to understand how people give shape to events, how they make a point, their reaction to events, and how they portray them. All of this can be used as a starting point for further exploration and analysis" (Gibbs 2007: 70). The authors put forward a structural model of narrative form based on fundamental techniques in linguistics, using mainly biographical interviews of telling personal experiences. The by now well-known model sets the first attempt to apply a linguistic approach to oral narratives of personal experience (Mishler 1995:

92; Elliott 2005: 42). Before the model is fully formally developed, the authors informally define narrative "as one method of recapitulating past experience by matching a verbal sequence of clauses to the sequence of events that actually occurred" (Labov and Waletzky 1997: 12). Developing the model further, the authors identify basic elements within a narrative: At the very base narratives consists of clauses which fulfill different functions. A narrative clause maintains the strict temporal sequence that is the defining characteristic of narrative. A free clause can range freely through the whole narrative. Furthermore there are coordinate clauses, that can be placed anywhere without disturbing the semantic interpretation or the temporal order, and restricted clauses, which are neither free nor temporally ordered. These clauses are defined in relation to the temporal sequence in the narrative and the resulting displacement set (i.e. a set of clauses that defines the temporal sequence). Taking the analysis to a higher level, the authors outline six basic components of simple narratives: Abstract, Orientation, Complicating Action, Evaluation, Result or Resolution, and Coda. The orientation section gives the audience information about places, persons, time, and behavioral situation the narrative refers to. Usually the larger body of narrative clauses covers several events that describe the complicating action. This section is closed by an evaluation given by the narrator where she comments on the events. "The evaluation of a narrative is defined by us as that part of the narrative that reveals the attitude of the narrator towards the narrative by emphasizing the relative importance of some narrative units as compared to others" (Labov and Waletzky 1997: 32). With the identification of the evaluation section, the section of the resolution becomes apparent. Here, the narrator draws a conclusion from the story she has told. The evaluation section and the resolution section can be given as one. Labov and Waletzky detect in their broad analysis of interview data that many narrators actually add a coda section to the end of the story. Those are clauses that do not contain information about the events, but function as closing remarks (in Labov and Waletzky's study they present phrases like "And that – that was it, you know", "And that was that", and "That was it" to be such closing remarks, Labov and Waletzky 1997: 36).

The following table demonstrates an application of the model to a section of one of this study's interviews. This section is from an interview with a painter in the river valley and was conducted in order to achieve an understanding of the meaning a place has for its community. In this section, the interviewee is asked to tell about her experience with an organized painting workshop in the river valley.

Exemplary analysis of clauses (derived from interview no. I-USA 3):

Orientation	68	Many youngsters were there. I was certainly the oldest
	69	student.
	70	They took me on the last minute, I sent in a couple of
	71	images,
	72	and they were meeting to wrap things up and they told […], the head of this thing, we definitely have to let this guy in. So we painted every day for about a month. We got up early and we would go out.
Complicating Action	74	I became close with probably two of the students.
	74–75	One was closer to my age, but that wasn't the reason. We
	76	painted similarly.
	78	We went on separate hikes together. I learned some technical skills.
Evaluation	79–81	But what I saw was that the students were very good at
	82–83	recreating what they saw, but the river…. You know, I am
	85–86	certainly not a very religious person, but the [school of old
	91–92	masters] was definitely tied up to religion.
	92–94	These men believed that they were walking into God's home when they were walking into this wilderness. And these younger people […] had a different view point. They would immediately plug in their iPods And I would listen, I would set up and I would be right near the stream of water. And I would be listening to the wood splashing, and to the birds, and everything around, taking it all in.
Resolution	96	I mean, it was fine…
	77	I kind of separated from that group.
Coda		– None –

Note that the resolution does not necessarily have to be stated at the end of the narratives. The statement "I kind of separated from that group" is given by the interviewee in the first section of the narrative sequence (line 77), but anticipates the result following the narrative clauses (lines 68 through 78) beforehand. This finding is in line with the assumption that even in biographical interviews "respondents rarely provide strictly chronological accounts (Elliott 2005: 46; Coffey and Atkinson 1996: 58).

The model helps to separate singular clauses and to uncover their function and relation to each other. This is particularly helpful in biographic interviews, with a high number of narrative sequences that correspond to the model's requirement of events being told. For narratives that do not refer to personal experiences, an application of the whole model does not prove fruitful. It becomes quite clear that applying this method to whole interviews is not very feasible (Elliott 2005: 46). The outlined section above should merely clarify the basic framework of the method. However, the approach by Labov and Waletzky provides a linguistic understanding of narrative structure, which helps to separate clauses also in less action driven narratives. Coffey and Atkinson advocate the model for it "provides us with an analytic perspective on two things: it allows us to see how that narrative is structured, and it offers a perspective from which to reflect on the functions of the story" (Coffey and Atkinson 1996: 61). The model thus guides researchers to avoid reading qualitative data purely for content by beginning with a formal analysis of the structure of a narrative (Elliott 2005: 42). If we take narratives to be social texts, as pointed out by Franzosi (1998: 517), than the analysis should not only cover telling of personal experience, but should also take comments and statements as narratives. Thus, from this linguistic approach I will draw the understanding of different functions of clauses in order to identify the borders of a narrative. With this I argue against Elliot, who criticizes the model to the effect that the various sets of narratives appearing in interviews cannot be separated that easily (Elliott 2005: 46). On the contrary, the model helps to identify boundaries between narratives, if the analyst takes into account that one clause can serve more than one narrative.

But for a sufficient analysis of narratives that understands their role in civil discourse, the analytic tool must include cultural and social elements.

The Structural Model of Genre: The Form of Narrative

The Structural Model of Genre laid out by Smith (2005) moves beyond the structure of single units as Labov and Waletzky do but does not concern itself with content-related topics, as McBeth and Jones do; rather, the model sets out to conceptualize narratives with the term of genres and analyze the implications of a genre guess. It considers arguments in cultural sociology claiming that narratives "just like binary codes, circulate and are contested in the collective conscience and in this process can shape history" (Alexander and Smith 2010: 17).

In his study on war narratives, Smith aims at developing an alternative explanation to the question, why nations engage in warfare apart from instrumental, material rationale like a struggle over power or resources.

With a cultural take on the issue, the author illustrates the importance of binary codes defining sacred and profane and a limited pool of narrative structure as cultural backdrop for legitimating military policy. In this chapter, I will portray Smith's structural model of genre and how it can be brought to use in the discussion of climate change narratives. The concept of genre is also discussed in Jacob's study on the events following the Rodney King Beating in 1991 as "an important part on how events get narrated, linked to other events, and infused with social expectation (Jacobs 1996: 1267).

Drawing on the description of civil discourse (Alexander and Smith 2003a) Smith differentiates two perspectives on analyzing meaning. Thought together, both axes are the foundations of the cultural system through which real world events as mere facts are turned into non-material social facts in the collective conscience: the paradigmatic approach understands meaning as brought about by binary codes which help classify the world we live in. Anthropological studies have emphasized the role of binary oppositions in so-called primitive societies, where myths and codes are used to construct collective representations of the world. These codes provide building blocks and this approach can be seen in the works of Lévi-Strauss, Barthes, and Sahlins. However, binary codes are yet sufficient for the explanation of cultural meaning. They often simplify the meaning of events, ignore subtle changes in the discourse and moreover they do not provide a toolbox for action. "Binaries help to make sense of the world, but they do not offer an instruction manual for what to do next" (Smith 2005: 17). To untangle the complexity of civil discourse, an analysis of narrative is needed.

This is provided by the second axis of the cultural system, the syntagmatic approach, which can be found in the works of Propp, Greimas, and Ricoeur and is centered in a diachronic and more sequential perspective. Narrative structures place actors and events into plots. They provide intricacy to our understanding of world events and convert situations into scenarios. Narratives order through sequencing rather than through distinction and resemblance (Ricoeur 1980: 171; 178) and thus create causalities within a story. Thus, narrative theory provides the link between culture and agency that is missing in the concept of binary codes. However, for Smith a generalizable theory of narrative is missing. For his structural model of genre, the author draws on hermeneutics, the idiographic and inductive approach to narrative in the social sciences, and structural analysis as shown in the works of Hayden White (1987). A cultural sociological approach to narrative, needs to draw on those schools: in the tradition of social sciences, it has to consider the role of narrative in real social life; in the tradition of hermeneutics, it has to take into account the implications narratives have for the ethical or normative regulation of social action; and in a structuralist tradition it has to aim for a generalizable theory that sees the limited canteen of cultural tropes that are narratives (Smith 2005: 19). Referring to Aristotle's poetics, Smith claims that narratives are limited as they consist of finite elements.

Analogue to Saussure's interest in langue rather than in parole, "structuralist poetics ignore the surface detail of this story or that drama in digging for the patterned relationships and regularities that unite particular genres of storytelling activity" (Smith 2005: 20). Those regularities and patterns are summed up and represented by specific genres, which give way to the outcome of a story through a genre guess. Frye (1969) develops a formalization of genre. This formalization focuses on the position and characteristics of the hero and thus comes to a hierarchy of fictional modes (genres).

> In literary fictions, the plot consists of somebody doing something. The somebody, if an individual, is the hero, and the something he does or fails to do, is what he can do, or could have done, on the level of the postulates made about him by the author and the consequent expectations of the audience. Fictions therefore maybe classified not morally, but by the hero's power of action, which may be greater than ours, less, or roughly the same. (Frye 1969: 33)

The hero in a story told in the Ironic Mode disposes of powers of action inferior to the audience's; Low Mimetic Mode presents the hero as "one of us" (Frye 1969: 34). In High Mimetic Mode, the hero is superior to other men, but not to his environment, whereas in romance and myth the hero is set apart both from the audience and nature. Myths tell "heroic stories of triumph over adversity" (Smith 2008: 91).

Smith argues that these genres – even though Frye is referring to fiction writing – also apply to the way in which we perceive real world events. Yet, Frye's typology of fictional modes must be advanced to be applicable to non-fictional narratives. Thus, Smith adds three characteristics to the hero's power of action: first, the shape of motivation: is it set in material interests or are the hero's concerns ideal? Second: the importance of the object of struggle, moving on a scale from local through national onto global. Third: the polarization between the hero and the villain: the larger the difference between both characters, the higher the genre.

Smith describes four genres occurring in the civil discourse of war:

> The low mimetic genre is the predominant narrative mode for the public understanding of everyday politics. It is little plot driven and emotionally flat. Smith describes it as the "genre of anti-history". (Smith 2005: 23)

> Tragedy provides a way to identify with the object of struggle that is often portrayed as innocent suffering. However, the genre fails at provoking action as fatalism is often a theme in this mode of storytelling. "The tragic genre can lead to paralysis in political life rather than demands for intervention". (Smith 2005: 25)

Contrary to the genre of tragedy, romance is indeed effective when it comes to demand action, as it is marked by the belief that actions can make a difference and a change for the better is possible. However, the genre proves to be highly unstable and vulnerable to attacks from advocates of realist and tragic visions, a characteristic that has comes from the suspicion that actions might end up being "futile and lead to unintended outcomes" (Smith 2005: 26).

The genre that drives engagement in conflict is the apocalyptic outlook on a situation. Due to a high polarization between hero and villain and the object of struggle being extremely significant, it is highly effective in "generating and legitimating massive society-wide sacrifice" (Smith 2005: 27) (Fig. 3.1).

The genres shown in the figure above are based on Frye's discussion of fictional mode, yet Smith develops genre types fitted to the civil discourse of war. He thus portrays the discussion of whether or not nations engage in military actions in specific situations as genre war. Influencing for a decision for or against war is, according to Smith, the genre guess that is inherent to each genre. That is, e.g., the apocalyptic genre evokes high level demonization of the villain, depicts the object of struggle as highest good and of global importance, and allocates idealistic motivation to the hero and a high belief in its powers to resolve the current situation to legitimate military action. This is demanded by the public, as a quote from Republican Pat Buchanan in the debate about the Gulf War of 1991 shows: "Before we send thousands of American soldiers to their deaths [...] let's make damn sure America's vital interests are threatened. [...] Saddam is not a madman, he is no Adolf Hitler" (Smith 2005: 112).

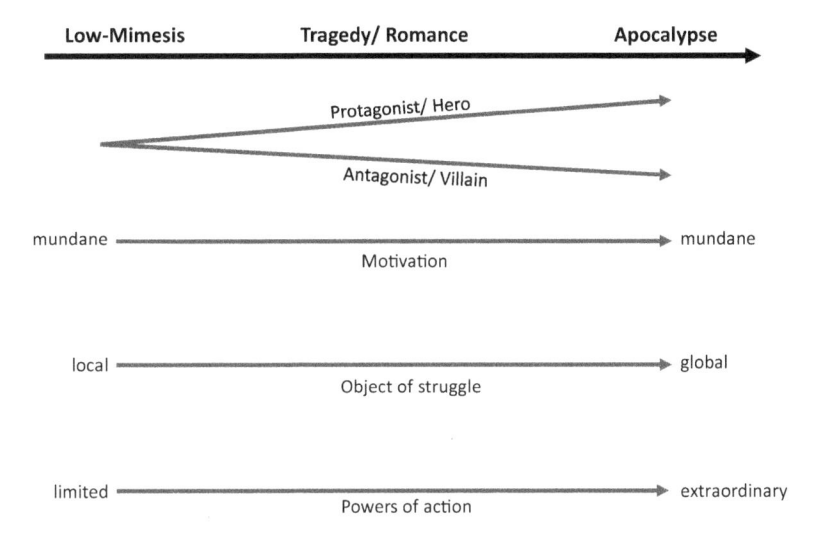

```
                Low-Mimesis         Tragedy/ Romance              Apocalypse
                                    Protagonist/ Hero
                                    Antagonist/ Villain
    mundane                                                                    mundane
                                         Motivation

    local                                                                      global
                                       Object of struggle

    limited                                                                    extraordinary
                                        Powers of action
```

Fig. 3.1 The structural model of genre (Source: Smith 2005: 24)

A reason affecting the people of the USA to engage in this war must be of high importance, otherwise military action will not be legitimated and accepted by the public. Applying the typology of genre to this case, Smith claims that the antiwar Right sought a low mimetic genre, downplaying the importance of this particular conflict to the American people and also downplaying the danger and risk that sprung from the villain Saddam Hussein.

Understanding the choice of a particular genre is not just an academic endeavor but actually tells a large deal about the reasons why a specific situation developed this way and not another. "Genre politics is a witnessable, reportable, measureable social fact that has determinate material consequences over multiple cases" (Smith 2005: 208). The genre that is chosen by an actor to tell a story works as a prediction for future events. "Genre influences the expected outcome of a particular narrative construction by constructing a set of expectations for the hero and the conclusion of the story. […Thus] a 'narrative sociology' can help social scientists to better understand the dynamics of social process and social change" (Jacobs 1996: 1267). This is what Smith depicts as genre guess (Smith 2005: 27–34): the evaluation of a situation based on the interpretation of a few events and then ongoing efforts to check this first interpretation as things develop.

At this point Smith's understanding of genre falls short. The author focuses on genre change within an actor-group solely over time; the time in which events develop. "It may be that we got our earlier readings wrong. […] We look back to the past, think about old clues in new ways, construct revised narratives, and contemplate changing our genre" (Smith 2005: 31). Besides the aspect of temporality – and especially if the narrative analysis is concerned with an explanation of social facts rather than fictional literature – genres change depending on the audience a specific actor group is targeting. I claim that there is not one way a story is told by a specific group, but that genres overlap can even be partly contradictory to one another.

THE NARRATIVE POLICY FRAMEWORK: THE CONTENT OF NARRATIVE

McBeth et al. (2005) argue that policy debates are less open to objective facts about an issue (if there is such a thing as objective facts) but that different solutions are supported by a differing examination of the situation and the inherent evidence, stemming from a selective interpretation of the circumstances at play. The concept of the social construction of the world leads also in policy studies to the question how meaning is developed and assigned. Narratives are key to understand this meaning making process as they are in an "epistemologically privileged position in making sense of a socially constructed world" (Jones and McBeth 2010: 334).

Thus, "examining policy controversies provides insights into competitive interest group framing" (McBeth et al. 2005: 419). Jones and McBeth (2010) develop a narrative policy framework based on Wildavsky and Douglas' Cultural Theory, which aims at satisfying positivist scientific standards by operationalizing narrative research paradigms. The authors apply a statistical approach to Narrative Policy Analysis (Roe 1994) by quantification of narratives and hypotheses testing (McBeth et al. 2005). Thus, this model focuses less on formal structures, as did Labov and Waletzky's model, but introduces cultural interpretation of events into the analysis of narratives. "The Narrative Policy Framework seeks to combine aspects of structuralist narrative analysis, where the emphasis is on textual elements, with a poststructuralist focus on subjective interpretation and the deconstruction of ideological agendas" (Paschen and Ison 2014: 5). Cultural Theory examines the social construction of meaning and the patterns with which individuals and groups interpret events that unfold in their world (Leiserowitz 2003: 58). Cultural Theory attempts to explore "the different perceptual screens through which people interpret or make sense of their world and the social relations that make particular visions of reality seem more or less plausible" (Schwarz and Thompson 1990: XIII). Portraying the debate between positivist and post-positivist scholars, McBeth and Jones argue for a policy analysis tool that matches positivist standards and can be "clear enough to be wrong" (Jones and McBeth 2010: 331).

Drawing on preceding works on narrative analysis in policy studies, the definition of narrative covers the basic elements: narrative thus must contain a setting or context, a plot, characters consisting of three general categories (heroes, villains, and victims), and a moral of the story, in this case a policy solution. For a statistical analysis, narrative content must avoid relativity, thus, "narratives must be anchored in generalizable content to limit variability" (Jones and McBeth 2010: 341). To derive such categories, the authors suggest turning to the concepts of partisanship and ideology and cultural theory as a scheme for belief systems.

Partisanship and ideology provide an understanding of characters, plot, and causal mechanisms that are used to describe a situation by opponents in political discourse. Loyalty towards one party acts as a cognitive filter (Bartels 2002, cited in Jones and McBeth 2010 341), ideology structures political preferences. If only Ideology is taken as basis for the analysis of the discourse around climate change, then the topic would be presented as a simply partisan issue. "The story told by ideology is straightforward – conservatives are much less likely to agree with the majority of scientists and, as such, are less likely to support climate change mediating policies. Liberals, by definition, are the exact opposite" (Jones 2011: 723). An Ideology based approach leaves a topic like climate change contentious and partisan, with no compromising middle ground to be found. Cultural Theory, the authors claim, opens the debate up to compromise among the different types and their perception of nature.

With a large body of literature that examines attitudes towards political issues, such as climate change, Cultural Theory has been proven to serve as a useful tool to identify culturally specific policy narratives, which allows the authors to use Cultural Theory as a robust anchor for narrative content. Thus, the analysis of narratives can be grounded in well-established theory.

From the viewpoint of sociology, another take on the debate between quantitative and qualitative methods and their respective justifications and scientific standards does not seem fruitful. However, for this study on climate change narratives, it is helpful to show that the study of narratives does not remain a nebulous and elusive concept, not fitted for underpinning political theory. Jones and McBeth attribute the reservations policy studies have over narrative analysis to the circumstance that narrative entered policy studies through the vehicle of post-structural literary theory. The quantitative, structuralist, and positivist approach does not seek to replace qualitative, post-structuralist analysis of narratives, but to enable positivist and post-positivist scholars to engage in fruitful debate over narrative's significance in shaping public's opinion (Jones and McBeth 2010: 339).

Shortcomings of the Presented Models of Narrative Analysis

The three analytic models focus on different elements of a narrative. Labov and Waletzky strip the narrative off its content to focus completely on linguistic units. This offers a way to understand the inner logic of a narrative and identify causalities in-between them. However, there are a few methodical and theoretical problems with it: firstly, as Elliot (Elliott 2005: 46) points out, applying this analysis tool to whole transcripts of interviews produces a huge amount of clauses that does not always prove useful for the analytic interpretation. Secondly, and more important for a social scientific approach to narratives: the model remains its analytic endeavor on the purely structural level. With this, a link to social theory in order to interpret and explain structural findings is missing.

On the other end of the hierarchy of structure, form, and content sits Jones and McBeth's narrative policy framework. Its basis of Cultural Theory proves to be very fruitful when it comes to explain various attitudes towards a social problem (in this case climate change). Yet, it imposes purely social categories to the data and ignores the potential that an analysis of textual structures bears for social explanation. A cultural sociological analysis should take advantage of its assumption to treat culture as text.

The model by Smith bridges the gap between those models that focus either on content or on structure by applying the theory of genres to narratives and thus making use of the genre guess concept. Smith's model, however, falls short on two accounts: firstly, a narrative has to fulfill each characteristic of the model on the same level to properly fit into a specific genre. There are, however, problems with some of the characteristics Smith proposes. Take for example the scale on which the object of struggle is evaluated. In order to fulfill the demands of the apocalyptic genre, the object of struggle has to claim global importance. But that ignores that civil discourse takes place on different levels, and that a narrative might have apocalyptic structures even if the object of struggle does not concern the whole planet. The model works well for the analysis of war narratives, because the consequence of an apocalyptic reading of events would always be global (most wars affect not only the nations, allegiances, religious, ethnic, etc. groups that are directly involved, but influence others as well). But if the model is to be applied to less high-stake topics, this scale has to be altered. Secondly, Smith allows a change in genre only as progressing over time, as more events are uncovered. He thus ignores a vital element of social discourse analysis: a topic can take form in different narratives with different genres and genre guesses within the same group of actors at the same time. This is especially true if one considers what Smith calls genre wars: consequences drawn from a list of events depend heavily on the reading, the narrative interpretation that succeeds in the public discourse. While this is true, Smith accounts only for one narrative per actor group. But this discourse is far more multi-layered than that.

The strongest point of criticism that goes for all three models is that all the models above fail to put different narratives in relation to each other. Smith goes for one singular narrative interpretation of one event, changing genre only over time and according to newly uncovered information. Jones and McBeth tie narratives back to theoretical ideal types, allowing each type only for one narrative understanding. Labov and Waletzky take the different elements of a narrative apart without implying an interpretation. But it is precisely when social facts are compared that we see their nature through the differences between them.

Structure – Form – Content: Towards an Integrated Model of Cultural Narrative Analysis

Taking this critical reflection into account, a model of narrative analysis which benefits from the above presented models must fulfill the following demands:

In response to Labov and Waletzky's model the structural analysis of an integrated model must not be autotelic, i.e. it should not pursue identifying structural elements of a narrative as an end in itself. Thus, we need to make sure that the analysis of those structural elements serves a socio-cultural purpose. Instead of focusing on purely textual clauses as Labov and Waletzky, I propose a more content-related approach by marking textual elements that focus on the active players within in the narrative: hero, villain, victim, and the relationship these elements keep with one another. Thus, for a social analysis of narratives the Structural Model displays exemplary the use of a sound textual analysis to identify singular elements without implying a hierarchical or judgmental interpretation.

Interpretational hierarchy should not overpower other analytic results. As Smith introduces a high demand for hierarchy in his analytic model, the identified elements need to meet each other's standard. That is that an analytic conflict occurs if, on the one hand, a violent contrast can be identified between hero and villain, and, on the other hand, the object of struggle is not a global threat. Smith's hierarchical scale for the object of struggle ignore that narratives are subjective to the point of view from which they are told. The scale of local – national – global imposes an objective, factual view on the object of struggle. But the assessment of its importance lies within the narrative, and is not externally imposed. Despite this, the concept of the genre guess introduces a means for a social analysis of action that a narrative evokes and thus carries the sociological purpose of the analysis.

As opposed to Jones and McBeth's use of Cultural Theory's strict categories, the proposed model will take advantages from literature theory's insights and will thus allow for a more open analytic approach. However, the strict use of the categories in the Narrative Policy Framework ensures that the model does not miss its purpose to serve an understanding of the social role of narratives, to comprehend the social background of a narrative, and not to pursue only purely literary and structuralist purposes. I draw from Jones and McBeth the possibility of identifying the main topic of a narrative, which can be found in general tropes of policy studies. Here, the storyteller and the audience can be identified, helping to carve out the relationship between these two actors of a narrative. The identification of these elements – storyteller – audience – main topic – is based on Polletta's claim that "the risks in storytelling come as much from the norms of narrative's use and interpretation as they do from the norms of its content. […] Stories are differently intelligible, useful, and authoritative, depending on who tells them, when, for what purpose, and in what setting" (Polletta 2006: 3).

The integrated model of cultural narrative analysis tackles two more issues neglected by the former approaches: firstly, it allows for the idea that multiple narratives exist within one social group, and that change in these narratives does not only occur over time and with the discovery of more facts, but that they differ depending on the social circumstances and the audience. The interplay between those narratives and their characteristics is pivotal to a social analysis as it provides insights of the structure of the public discourse. A comparison of different narratives that have emerged in the analysis thus promises to reveal how these different stories influence one another and which consequences possible interdependencies have for their stability and reliability.

The model follows Hayden White and Ferdinand de Saussure with their differentiations between content and form resp. parole and langue by segregating three levels of analysis (Fig. 3.2):

The single circles of the model can be further explained on their own:

Structure The circle on the right commits to the identification of structural elements in the story and thus draws intensely on narrative theory and literature theory. Singling out the clauses that describe either the hero, the villain, or the victim, helps to uncover implicit references to certain characters. It also allows for an understanding of the relationships between those elements. As I will show later in the analysis the explicitly mentioned characters are as important to the model as those that are omitted. This paradigmatic analysis of narratives makes use of structural binary codes by placing the actors in patterns of opposition. This analytic step only becomes possible by allowing for a comparison between the narratives.

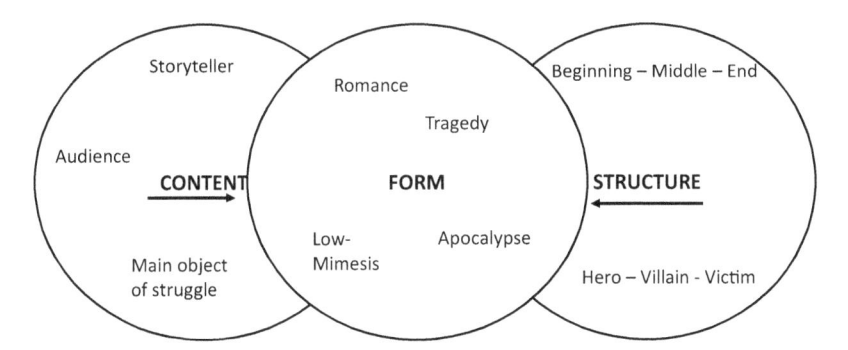

Fig. 3.2 Integrated model of cultural narrative analysis (own illustration)

Content The circle on the left in the model combines the broader context of a story as it is framed in the public discourse, resembling Saussure's concept of parole, which he describes as speech as "an individual act of the will and the intelligence", and – with a little license – is expandable to the sense of discourse as is presented by the German Rede (Saussure 1986: 16). This circle represents the semantic topic of the narrative and draws on the analytic findings in the circle on the right (Structure) to describe the point of view of the storyteller (narrator) and the audience she addresses. The focusing on the storyteller allows to identify the main topic of the narrative, or the object of struggle, with recourse to political, cultural, and social tropes.

Form The circle in the middle focuses on the form of a narrative. Drawing on Smith who follows Aristotle's classification of literary genres, this part of the model values the benefits of the genre guess. Narratives are thus a sense-making cultural form, that links actions to cognitive and value systems. The encompassing concept of genre enables the model to bring the different elements back together to form a cohesive picture of the story that is told. The choice of genre depends on both the lower level of Structure and the upper level of Content. The relationship between the elements and the characteristics of those elements as described and identified on the level of Structure determines the literary form the narrative takes in the public discourse.

At the center of the model lies a basic understanding of culture as text as it is basic to cultural sociology. The interpretation of social facts is analyzed with a combination of social, cultural, and literature theory, allowing for a holistic analysis of the underlying cultural patterns within social interpretation. With cultural sociological methods we will be able to analyze climate change narratives as they are brought in use by climate change advocates. We will understand their cultural meaning and uncover connections and inconsistencies that hamper successful climate change communication.

NOTES

1. Ryan underlines the importance of causality for a definition by citing e.g. Onega and Landa ("The semiotic representation of a sequence of events, meaningfully connected in a temporal and causal way." – Onega Jaén and García Landa (1996): 3) and Bal ("The transition from one state to another state, caused or experienced by actors." – Bal (1997): 182). Roland Barthes

((1975): 271) states that with people's efforts in telling their experience of the world in language, narratives establish meaning where before was a mere copy of the events that are told.
2. Ryan refers to Ricoeur („I take temporality to be that structure of existence that reaches language in narrativity, and narrativity to be the language structure that has temporality as its ultimate reference." – Ricoeur (1980): 165) and Brooks ("Plot is the principal ordering force of those meanings that we try to wrest from human temporality." – Brooks (1992): ix).

REFERENCES

Abbott, H. P. (2007). Story, Plot, and Narration. In D. Herman (Ed.), *The Cambridge Companion to Narrative* (pp. 39–51). Cambridge: Cambridge University Press.

Alexander, J. C. (2003). On the Social Construction of Moral Universals: The "Holocaust" from War Crime to Trauma Drama. In J. C. Alexander (Ed.), *The Meanings of Social Life. A Cultural Sociology* (pp. 27–84). Oxford/New York: Oxford University Press.

Alexander, J. C., & Smith, P. (2003a). The Discourse of American Society. In J. C. Alexander (Ed.), *The Meanings of Social Life. A Cultural Sociology* (pp. 121–155). Oxford/New York: Oxford University Press.

Alexander, J. C., & Smith, P. (2003b). The Strong Program in Cultural Sociology: Elements of a Structural Hermeneutics. In J. C. Alexander (Ed.), *The Meanings of Social Life. A cultural sociology* (pp. 11–26). Oxford/New York: Oxford University Press.

Alexander, J. C., & Smith, P. (2010). The Strong Program. Origins, Achievements, and Prospects. In R. R. Hall, L. Grindstaff, & M.-C. Lo (Eds.), *Handbook of Cultural Sociology*. London/New York: Routledge.

Bal, M. (1997). *Narratology: Introduction to the Theory of Narrative* (2nd ed.). Toronto/Buffalo: University of Toronto Press.

Bartels, L. M. (2002). Beyond the Running Tally: Partisan Bias in Political Perceptions. *Political Behavior, 24*, 117–150.

Barthes, R. (1975). An Introduction to the Structural Analysis of Narratives. *New Literary History, 6*(2), 237–272.

Boholm, Å. (2015). *Anthropology and Risk. Earthscan Risk in Society Series*. London: Routledge, Taylor & Francis Group.

Bonnell, V. E., & Hunt, L. (Eds.). (1999a). *Beyond the Cultural Turn: New Directions in the Study of Society and Culture*. Berkeley/Los Angeles/London: University of California Press.

Bonnell, V. E., & Hunt, L. (1999b). Introduction. In V. E. Bonnell & L. Hunt (Eds.), *Beyond the Cultural Turn. New Directions in the Study of Society and Culture* (pp. 1–31). Berkeley/Los Angeles/London: University of California Press.

Brooks, P. (1992). *Reading for the Plot: Design and Intention in Narrative*. Cambridge, MA: Harvard University Press.
Bruner, J. (1991). The Narrative Construction of Reality. *Critical Inquiry, 18*(1), 1–21.
Coffey, A., & Atkinson, P. (1996). *Making Sense of Qualitative Data. Complementary Research Strategies*. Thousand Oaks: Sage Publications.
Cortazzi, M. (1993). *Narrative Analysis*. London/Washington DC: The Falmer Press.
Dahlstrom, M. F. (2010). The Role of Causality in Information Acceptance in Narratives: An Example from Science Communication. *Communication Research, 37*, 857–875.
Elliott, J. (2005). *Using Narrative in Social Research: Qualitative and Quantitative Approaches*. London/Thousand Oaks: Sage Publications.
Franzosi, R. (1998). Narrative Analysis – or Why (and How) Sociologists Should Be Interested in Narrative. *Annual Review of Sociology, 24*, 517–554.
Frye, N. (1969). *Anatomy of Criticism: Four Essays*. Atheneum/New York: Princeton University Press.
Gibbs, G. (2007). *Analyzing Qualitative Data. SAGE Qualitative Research Kit*. Los Angeles: Sage Publications.
Halttunen, K. (1999). Cultural History and the Challenge of Narrativity. In V. E. Bonnell & L. Hunt (Eds.), *Beyond the Cultural Turn. New Directions in the Study of Society and Culture* (pp. 165–181). Berkeley/ Los Angeles/ London: University of California Press.
Hards, S. (2012). Tales of Transformation: The Potential of a Narrative Approach to Pro-environmental Practices. *Geoforum, 43*(4), 760–771. https://doi.org/10.1016/j.geoforum.2012.01.004.
Hinchman, L. P., & Hinchman, S. (1997). Memory, Identity, Community: The Idea of Narrative in the Human Sciences. In *SUNY Series in the Philosophy of the Social Sciences*. Albany: State University of New York Press.
Jacobs, R. N. (1996). Civil Society and Crisis: Culture, Discourse, and the Rodney King Beating. *American Journal of Sociology, 101*(5), 1238–1272.
Jones, M. D. (2011). Leading the Way to Compromise? Cultural Theory and Climate Change Opinion. *PS: Political Science and Politics, 44*(4), 720–725.
Jones, M. D., & McBeth, M. K. (2010). A Narrative Policy Framework: Clear Enough to Be Wrong? *Policy Studies Journal, 38*(2), 329–353.
Labov, W., & Waletzky, J. (1997). Narrative Analysis: Oral Versions of Personal Experience. *Journal of Narrative & Life History, 7*(1–4), 3–38.
Leiserowitz, A. A. (2003). *Global Warming in the American Mind: The Roles of Affect, Imagery, and Worldviews in Risk Perception, Policy Preferences, and Behavior* (Unpublished Dissertation). University of Oregon, Eugene. Retrieved from http://decisionresearch.org/pdf/540.pdf
Margolin, U. (2007). Character. In D. Herman (Ed.), *The Cambridge Companion to Narrative* (pp. 67–79). Cambridge: Cambridge University Press.

McBeth, M. K., Shanahan, E. A., & Jones, M. D. (2005). The Science of Storytelling: Measuring Policy Beliefs in Greater Yellowstone. *Society & Natural Resources*, *18*(5), 413–429. https://doi.org/10.1080/08941920590924765.
Mishler, E. G. (1995). Models of Narrative Analysis: A Typology. *Journal of Narrative and Life History*, *5*(2), 87–123.
Onega Jaén, S., & García Landa, J. Á. (Eds.). (1996). *Narratology: An Introduction, Longman critical readers*. London/New York: Longman.
Paschen, J.-A., & Ison, R. (2014). Narrative Research in Climate Change Adaptation – Exploring a Complementary Paradigm for Research and Governance. *Research Policy*, *43*(6), 1083–1092.
Plummer, K. (1995). *Telling Sexual Stories: Power, Change and Social Worlds*. London: Routledge.
Polletta, F. (2006). *It Was Like a Fever: Storytelling in Protest and Politics*. Chicago/London: University of Chicago Press.
Richardson, B. (2007). Drama and Narrative. In D. Herman (Ed.), *The Cambridge Companion to Narrative* (pp. 143–155). Cambridge: Cambridge University Press.
Richardson, L. (1990a). Narrative and Sociology. *Journal of Contemporary Ethnography*, *19*(1), 116–135.
Richardson, L. (1990b). *Writing Strategies: Reaching Diverse Audiences, Qualitative Research Methods* (Vol. 21). Newbury Park: Sage Publications.
Ricoeur, P. (1980). Narrative Time. *Critical Inquiry*, *7*(1), 169–190.
Roe, E. (1994). *Narrative Policy Analysis: Theory and Practice*. Durham: Duke University Press.
Ryan, M.-L. (2007). Toward a Definition of Narrative. In D. Herman (Ed.), *The Cambridge Companion to Narrative* (pp. 22–35). Cambridge: Cambridge University Press.
Saussure, F. D. (1986). *Course in General Linguistics*. La Salle, Illionois: Open Court.
Schwarz, M., & Thompson, M. (1990). *Divided We Stand: Redefining Politics, Technology, and Social Choice*. Philadelphia: University of Pennsylvania Press.
Smith, P. (2005). *Why War? The Cultural Logic of Iraq, the Gulf War, and Suez*. Chicago: University of Chicago Press.
Smith, B. (2007). The State of the Art in Narrative Inquiry: Some Reflections. *Narrative Inquiry*, *17*(2), 391–398.
Smith, P. (2008). *Punishment and Culture*. Chicago: University of Chicago Press.
Smith, T. (2010). Discourse and Narrative. In R. Hall, L. Grindstaff, & M.-C. Lo (Eds.), *Handbook of Cultural Sociology* (pp. 129–138). London/New York: Routledge.
Stone, D. (2002). *Policy Paradox: The Art of Political Decision Making* (3rd ed.). New York: W.W. Norton.
White, H. (1987). *The Content of the Form: Narrative Discourse and Historical Representation*. London: Johns Hopkins University Press.

CHAPTER 4

Telling the Stories of Climate Change: Structure and Content

Abstract This chapter presents the interpretation and analysis of empirical findings from qualitative data. Interviews were conducted with German and US-American climate advocates, i.e. people who are committed – professionally or as volunteers – to combat climate change and to get others engaged in the cause. Different narratives are identified and analyzed according to their structure and content, resulting in five main narratives derived from the empirical data. Here, economic concerns play a role, as well as environmental concerns and the way in which interviewees perceive the role of their government and country. Preserving nature and global solidarity with those who are suffering the impacts of climate change are put forward as values on their own, which justify getting involved in the fight against climate change.

Keywords Qualitative interview • Qualitative data analysis • Narrative analysis • Climate change • Environmental activism • USA • Germany

The following chapter will present the findings of the empirical analysis with regards to the theoretical framework developed in Chap. 3. Each identified narrative is described first by its content, including topic, storyteller, and audience. Second, the structure of the narrative – consisting of its characters – hero, villain, and victim, is presented.

© The Author(s) 2018
A. Arnold, *Climate Change and Storytelling*,
Palgrave Studies in Environmental Sociology and Policy,
https://doi.org/10.1007/978-3-319-69383-5_4

Methods

Data collection began in 2010 in Germany and continued in 2012 in the USA. For this study, 15 narrative, problem-centered interviews were conducted with 17 interviewees. Participants recommended people they encountered professionally or on a private level, e.g. through neighborhood or through climate change related activities within their communities. Participants were ensured anonymity in the beginning of the interview. It was also made clear in the beginning to the participants that professional expertise as well as personal should be equally voiced.

Interview no.	Occupation
I-GER 1	Subject specialist sustainability and climate, regional assembly
I-GER 2	Regional climate mitigation officer
I-GER 3	Member of state parliament, priority: energy and sustainability
I-GER 4	Journalist, regional newspaper
I-GER 5	manager regional assembly
I-GER 6	Regional head of law and administration
I-GER 7	Executive member regional flood aid group
I-GER 8	Regional director environmental NGO, Germany
I-USA 1	Reference person to an environment protection project
I-USA 2.1	Coordinator environment protection project
I-USA 2.2	Coordinator environment protection project
I-USA 3	Artist
I-USA 4	Artist
I-USA 5	Mayor
I-USA 6	Geophysicist (retired)
I-USA 7	Environmental activist

Not only was the occupation or the involvement of the respondents critical to the sampling process but there was also a regional scope introduced: before choosing and contacting possible interviewees, a literature research on afflicted regions was conducted beforehand. The concept behind this was that the study should not be limited to the perspective of people who are professionally involved in climate mitigation or related environmental activities but that it should also include the interpretation of events by laypeople. This approach was set to widen the scope of the study and to provide the possibility to match those narratives given by professionally involved personal with those of afflicted citizens. The regional scope thus was on the one hand to limit the available data, and on the other hand to ensure that interviewees shared a common background.

This way, the data was not flooded with narratives in reference to particular geographic environment that would not be present in any other interview. Additionally, the selected regions should not be existentially threatened by the consequences of climate change to avoid a circumstantial bias in the dataset. At the same time, the region should have encountered the changes that come with a changing climate on a quite moderate level so that it was possible to find concerned citizens that would not only mirror narratives presented in the media. The vulnerability of these regions is only one reason for their selection. Criteria for selection come from epistemological and pragmatic concerns, referring not only to the detailed regional scope of the cases, but more general to the selection of a region in Germany and in the USA.

Both regions were experiencing flooding, either from heavy precipitation or from rising sea level. The first interviews were set in the Germany. The region is defined by a river running through it. This region experienced some severe flooding in the recent past and increasing heavy precipitation during the winter months and heatwaves during summer are of concern to the region. The second case study is set in a region in the US. Interviews here took place between September 2011 and April 2012. This region is characterized by a larger river, houses in the town that is the main focus here, are often close to this river. Both case studies were narrowed down to one vulnerable town each. For anonymity's sake, I will refer to the US American case study as "Little Town" and to the German case as "Kleinhausen".[1]

Both regions are labeled as case studies, however, there is no case comparison intended in this study. On this level of politically active or at least interested climate advocates a common ground in the understanding of the issue is established. Discourse on the issue of climate change takes place within an area we might call 'expert sphere', with activists, politicians, and concerned citizens alike applying similar arguments and logic. For the study, it is interesting to see how narratives in the U.S. American interviews and the German interviews resemble each other. Interviewees move in the same circles and read similar reports and documents on the ecological crisis. Therefore, akin thought patterns emerge. Interviews in two globally different regions were conducted to avoid the possibility of a 'national bias', i.e. a nationally narrow reading of climate change and to get a wide range of narratives. In accordance with a thick description approach: "You can study different things in different places [...] but that doesn't make the place what it is you are studying" (Geertz 1973: 22).

The overall arching theme of all presented narratives is the topic of climate change. However, each narrative takes a different approach to the issue. Thus, four narratives were identified. These stories are not independent entities; they intersect with each other and draw on the same pool of characters. Thus, one statement or paragraph in an interview can play different role in different narratives, depending on the context. Also, the position of the characters varies in each story, thus leading to a varying hierarchical order between the characters (Fig. 4.1).

The overview of these narratives shows how climate change touches various spheres of social life. Renn assigns the amplification of climate change as a serious threat to the fact that "the impacts resonate with the concerns that are linked to each functional system of society" (Renn 2011: 164). This analysis can be seen in the figure above: climate change concerns the fields of economy (monetary loss in Renn's analysis), as well as social solidarity and questions of the relationship between humans and nature (social cohesion and environmental justice).

The first part of every chapter presents the content, focusing on the main topic (e.g. climate change in relation to economic development), identifying the characterization of the storyteller (e.g. in the narrative focusing on economic development, the storyteller might present herself as a rational, economically driven person). This again leads to a characterization of an

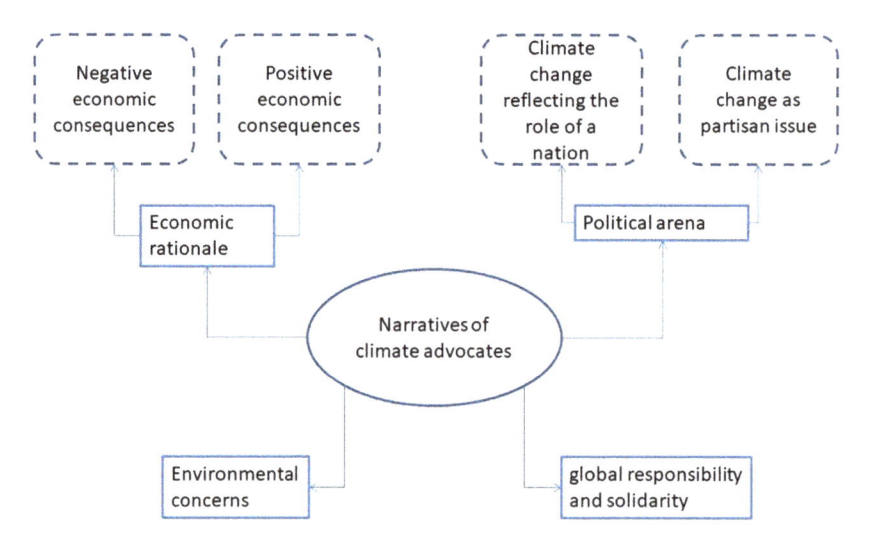

Fig. 4.1 Climate advocates' narratives (own illustration)

imagined audience. This means that the storyteller chooses and presents her arguments for fighting climate change according to the audience they are addressing. Again, in the economic narrative this audience is depicted as financially conscious and worried about the economic outcome of environmental involvement.

The second part of each chapter focuses on the structure, thus characterizing the villain, the victim, and the hero. Here it is important to note, that these characterizations change between narratives on the one hand. However, they also can change within a narrative, allowing for example a villain to become a hero, if she decides to get involved in the cause, advocated by the hero.

Putting a Price on Things: Serving the Economic Rationale

The first narrative I will discuss in length is the one that depicts climate change as an economic issue. Here, interviewees introduce environmental concern as a way to enhance the economic situation, i.e. investing in renewable energies and creating jobs in the process. This narrative can best be summed up with a quote from one interviewee, stating, that "We [the US] could be setting ourselves up to be a leader in renewable energy and adaptation strategies and we are not taking advantage of the biggest opportunity we have."

The economic rationale plays both ways: on the one hand, advocates describe the huge opportunities that the development of sustainable economics would provide. Thus, combating climate change becomes framed as positive economic challenges, country's economies could benefit from.

On the other hand, the dire consequences of climate change for the economic development are also described in a negative economic narrative. E.g. advocates stress the dangers a rising sea level or more and more frequently occurring heavy rainfalls may have on the manufacturing industry or how dwindling snow reliability in skiing areas can impact the tourism sector.

The narrative of economy deals with the economic effects climate change might have. As we will see in the following, climate advocates make use of both economic incentives (positive economic consequences) and financial losses (negative economic consequences). This can also be seen in the story of price Ney and Thompson (2000) identify in climate change talk. Here, the free market will achieve a reduction in consumption by introducing financial incentives.

Negative Economic Consequences

Content

The topic in this narrative resolves around negative impacts of climate change to national economies. Climate advocates emphasize that global warming will have negative economic consequences. Hereby they explicitly refer to the economy in industrialized countries.

In the German case study, interviewees describe the threats to the tourism sector in their region. These threats are already being experienced:

> Consequences in the tourism sector, if you think about winter tourism that is something we have to expect. That is well known already, [...]. The last two winters were quite alright, but before that, then we had ski lifts that weren't utilized to the optimum level, or that weren't running at all. (I-GER 5: 15)[2]

But interviewees also emphasize the future character of those threats:

> In the long run, we will most likely have less snow. And this will of course hit the tourism sector, the northern region around here with its smaller skiing areas. (I-GER 2: 6)

Interviewees in Little Town also hint at the damage climate change might do to their town's economy:

> You have to understand too, economically the... – in a sense, this is a perfect storm. The towns along the river have been renovating and developing incredible economic power over the past 15+ years. That is because the local river is getting cleaner. Consequently, at the same time we were beginning to see climate change and all these other issues impacting the river. While these towns had been developing – or I should not say developing, re-developing, re-inventing themselves, it has really been in the face of now. Water that is here, tides that are higher, and flooding that will continue to happen. (I-USA 5: 18)

After leaving its history as an industrial town behind, Little Town developed into a recreational environment for citizens escaping the bigger cities on the weekends:

> And in the summertime or in the fall you can't drive down this road because there are so many people coming in from the city, you know, apple picking, to look at the foliage […]. (I-USA 2: 55)

Severe flood damage will threaten this line of commerce.

Besides these threats to the tourism sector, forestry and agriculture are suffering from long periods of drought, both in the summer and in the spring, which is even worse for farmers:

> We have to expect longer and hotter summers. And in the region around the northern Black Forest you can see different phenomena happen already. Consequences in the timber industry […] Trees that are typical for this region, for example spruces, will be endangered. (I-GER 5; 15)

> I am just writing an article on that issue, farmers in the region denote that this spring drought right now is the worst they've ever experienced. A spring drought is the worse than a drought period during summer, because most of the harvesting is done by then. (I-GER 4: 24)

Longer drought periods will cause poor harvest due to a lack of precipitation:

> At the moment missing precipitation is hurting the agriculture and everyone is envisaging poor harvest this year. (I-GER 3: 24)

Companies which manufacture along the river are also affected by heavy precipitation, as one interviewee, a subject specialist in sustainability and climate, points out:

> When you think about the automotive industry for example, they need certain parts at a specific time, and when streets are flooded and the infrastructure breaks down, the whole supply chain gets interrupted. (I-GER 1: 14)

The perspective of interviewees in Little Town, USA, involves less direct economic losses like those described above, but emphasizes how rising sea level will affect the town's budget. For instance, it will threaten its tax base in the long run, when homes and buildings along the waterfront cannot be sustained anymore:

It is not just the buildings, which of course are a tax base for the village, so it has to be concerned whether it can maintain this tax basis, therefore invest into the infrastructure to maintain that tax base, or whether in the long run it will lose that waterfront as a tax base, so to give up on it. (I-USA 6: 16)

Adaptation measures that have to be taken to control the damage flooding can do are also a financial challenge for the town:

We just completed this project, which for us as a village was very expensive, putting these walls up in the sea level parking lot so the water can escape when it comes in on high tides. And it was a big project for us, and it won't work for long, but for now it is protecting parts of our village from being under water […]. (I-USA 4: 14)

This argument about economic threats is addressed to the financial conscious of fellow citizens. One interviewee, who has worked at a climate research institute and is therefore in demand as an expert on the subject in his hometown, puts the economic reason at the center of his argument:

I myself often fall into the mode of being 'Mr. Doom' instead of engaging them [audience] in positive acts. 'Mr. Doom' by scaring people too much with fore coming disasters.[3] While that helps on one level, particularly when it comes to the financial loss estimation, I always insist that whatever I do, I put a dollar sign on it. So, when I say what is the probability of such and such to occur, I say well it will cost the village a million dollars per event – or whatever. (I-USA 6: 28)[4]

Climate advocates employed by environmental NGOs follow this argument by also pointing out the costs to local government:

We are taking it home to the people who we think are gonna be most affected. For example municipalities that sit around the river shoreline are affected by sea level rise and if they don't think forward about it they will be reacting when the damage is done. So we try to reach out to them. (I-USA 2: 29)

Local officials pass this concern on to the citizens of their towns, explaining that changes will happen and that homeowners will have to deal with them. Asked about possible restrictions when homeowners seek to do major renovations on buildings that are set in flood zones, a local official replied:

I don't think I would be faced with legislating or restricting development at the river's edge. I would say, by all means, go ahead! But understand that this is the reality this is what is happening and this is how it is going to be like 50 to 100 years from now. (I-USA 5: 16)

These quotes show how the content of a narrative of negative economic consequences developing around the threat of financial losses (decline in tourism, crop losses, expenses due to flood adaptation). The storyteller depicts herself as a rational, economically thinking person, thus trying to persuade her audience with hard facts. By emphasizing the financial threats, climate advocates try to propose an argument which brings climate change closer to their audience, as it is a direct impact on their lives.

Structure

Like in other narratives, heroes, villains and victims are not mutually exclusive. Heroes in this narrative are those actors that work within the economic and political system. Governmental entities depict themselves as providing a framework for climate mitigation:

The economy can't handle this. But we as a government, we as a collective are here to put things in order. (I-GER 3: 38)

They emphasize the challenges they took on and point out how political decisions can prevent greater damages to citizens and the economy:

So, we decided to go another way und applied for the European Energy Award. I do believe that we can actually achieve more with this. We can develop a concept and then launch this concept, even if it takes some time. I believe this is the right way: to implement the goals of climate mitigation in our own concept, then reach those goals and even maybe exceed them. (I-GER 2: 36)

At first, there was some reluctance but then there was a good acceptance [of flood zone maps]. We said: We can't designate an area if it isn't protected. If people start constructing there they will basically lynch us and ask: How could you designate this area for building land if you knew there were other areas more suitable? (I-GER 3: 18)

The last quote already indicates one of the depicted victim figures in this narrative: property owners. One interviewee who also owns a home in a flood zone explained about adaptation measures, that he "simply paid an amount of 12.000 U.S. American Dollars to raise my house in 2003 when we bought this piece of property" (I-USA 6: 12). This is validated by a local official stating that "property owners have to take a lot of that [i.e. costs for adaptation measures] on" (I-USA 5: 12). Interviewees point out that not only immediate flood damage or adaptation measures debit property owners but that flood events can entail further financial challenges:

> All the houses are basically restored by now, but the flood has dramatic repercussions for large parts of the town population: All property owners who were hit by the flood can't get flood insurance anymore – or only with highly increased insurance contribution. (I-GER 3: 10)

As noted above as topic of the narrative, farmers and the tourism sector are described as victims of climate change on an economic level.

It is worth noticing that not only local challenges are mentioned within the realm of economic consequences, as this interviewee shows:

> If you take New Orleans[5] for example. I mean if that is also a sign of climate change… Well, we don't have to expect those kinds of devastating effects here. I mean, the losses there literally are going into millions and almost a whole city was destroyed. (I-GER 5: 54)

Interestingly enough the storytellers are not blaming any specific group of actors. Analyzing the structure of the narrative the only party that is found guilty are those impacts of climate change. However, as these are not actors in a sociological sense, but merely factors, the position of the villain in this story is left blank. Narratives are not self-contained, they interact with each other. In most narratives, industrialized nations or ruthless companies are depicted as villain in the story of climate change; in this presented narrative, they are presented as victims, as being vulnerable to impacts of climate change. Thus, they cannot – at the same time – serve as villains.

Positive Economic Consequences

Content

Contrary to the narrative where financial losses are at the center, climate advocates also make use of the argument that fighting climate change will have enormous economic benefits. This is very persuasive described in the following quote:

> I think the only way we can get through the challenge of the politicization of climate change is if we can show people a direct benefit, either in a positive economic development, or recreational development or some way to create jobs. If we can't connect this, because that is what people respond to, esp. with the economy the way it is. I think, that's the best way, to try to get our message across. Trying to connect what we are recommending in terms of solutions. Not talk so much about the problem, but saying: you all know, you're all seeing what is happening and these are ways to address it and then have the solutions be connected so that they have a lot of co-benefits for the community. And I think they won't address it for climate change's sake, for something that is years from now. They will address it if it is like, "oh, yes, if we design it that way, maybe we won't have the community, or the road flooded anymore." these co-benefits. (I-USA 2: 40)

Here it shows how the value of nature itself is pushed into the background by the prospect of economic opportunities; note how the interviewee in the quote above emphasizes that people "won't address it for climate change's sake", or that "a lot of the changes that have been made in the recent past were driven by economic interests. That had nothing to do with climate change." (I-GER-7: 58).

The argument of economic opportunity is made for local communities as well as on the level of national economy:

> And then there are partners like the nature conservancy […] they're all trying to figure out how to help the towns around here to protect their scenery. And protect habitat. Because they see this as a huge economic benefit for them. (I-USA 2: 55)

> We are missing a huge opportunity to be America first. We could be setting ourselves up to be a leader in renewable energy and adaptation strategies and we are not taking advantage of the biggest opportunity we have. (I-USA 2: 35)

The possibility of co-benefits is emphasized and politics are brought in as means to provide a framework which allows taking advantage of these economic opportunities:

Well, if you really think about it, it is a win-win-situation. If I approach the issue of energy saving, I firstly save money and secondly I get a climate mitigating effect. (I-GER 2: 18)

And incentives need to be part of the political agenda, for example via the KfW.[6] If people do not have the financial resources to invest in renovations one has to offer loans with reduced rates of interests. And that will benefit the national economy. (I-GER 1: 22)

The argument is clearly addressed at an economically driven audience which might also be under financial pressure such as local politicians:

I think the mayors yet to realize that this actually benefits them and their communities in multiple ways. (I-GER 2: 18)

Or if there are incentives they can take advantage of, to understand the... you know, when you start looking at the payback for different energy efficiency measures it becomes very clear to a local official that they could save some money. (I-USA 2: 40)

Interviewees express directly how this argument is specifically constructed to persuade this type of audience:

If we cannot connect this, because that is what people respond to, especially with the economy the way it is. I think, that's the best way, to try to get our message across. [...] Saying: You all know, you're all seeing what is happening and these are ways to address it and then have the solutions to be connected so that they have a lot of co-benefits for the community. (I-USA 2: 40)

The storyteller tries to achieve congruence with the audience in order to present themselves not as accusing, but as part of the challenge. Note how the interviewee in I-USA 2 emphasizes that "WE are missing a huge opportunity to be America first". Or how the interviewee in I-GER 2 addresses her audience by saying "if YOU really think about it, it is a win-win-situation" before bringing the argument into a rather active form: "if I approach the issue of energy saving, I firstly save money and secondly I get a climate mitigating effect". This narrative of a positive economic message follows Liverman's (2009) observation of a proposed market solution to the environmental crisis: carbon trading was introduced as a market instrument to manage pollution and led the way to climate mitigation as an investment opportunity.

Structure

Concerning the structural elements, climate advocates see actors within the economic realm as in the character of heroes, even though they are not acting accordingly just yet. The structural logic is: if these actors actually seek economic opportunities they will turn into heroes. As one interviewee put it:

> Well, we are not leaders. America is very America first, even though we are missing a huge economic opportunity to be America first. We could be setting ourselves up to be a leader in renewable energy and adaptation strategies. (I-USA 2: 35)

In this set up America has yet to take on this role of a hero in their own economic interest, they have not owned up to these challenges so far, but "we could be setting ourselves up to be a leader in renewable energies" implies that there is a possibility for the nation to reframe parts of their economy and contribute to stopping global warming at the same time. Forethought, innovation, and pioneering are characteristics of a hero in industrial setting:

> Here in Germany and especially in this state we were always at the developmental forefront, because we were pioneers. Between 1988/89 and 1995 I worked for a car company as an engineer. And we had an assignment to develop a car for a new global market, affordable for everyone. [....] As a young engineer I said that we can't "motorize" 1.4 million people. I literally said: The earth won't stand that! [...] And I think that at least our economy knows that we have to take on a pioneering role and lead this development. (I-GER 3: 30)

However, this narrative does not only depict national economies as entities to be actors in the fight against climate change, but employs also the financial elite of a society to advance technological innovations:

> Social innovations are always costly in the beginning and not necessarily economically and financially appealing. So you need wealthy people for that. Innovative technologies should be implemented with wealthy people, and afterwards step-by-step with the rest of the population. [...] If people with financial resources are involved and a product is forced into the market, prices will drop and the technology will become way more interesting. (I-GER 3: 30)

Again, the position of the villain is omitted in this narrative. Compared to the other narratives of climate change presented in this study, it seems that global economy and national economic interest are usually blamed, either as perpetrator of climate change or as perpetrators of failed global negotiations for a binding climate treaty.

This narrative stands out also by another factor: talk about economic chances is not set in relation to victims of climate change. The realm of economic and financial rationale does not allow for a victim figure. Here, climate advocates adopt the rationale of the economic world, where profit and money rules.

CLIMATE CHANGE IN THE POLITICAL ARENA: TWO CONTEXTS OF CLIMATE CHANGE AS A POLITICAL ISSUE

The first narrative in this chapter on climate change as a political topic explores the character setting behind the partisan battle over climate change. This narrative is specific to the US, but is referenced in the German interview data as well. The partly opposed views on the subject create a political environment, in which climate change is no longer a global crisis that has to be dealt with, but is just one more topic in the arena of partisanship and advancing a political ideology.

After reflecting on the role climate change plays in domestic politics, esp. in the US, the second narrative places climate change on the global map, describing how the topic carries different ideas of a nation's role within global politics and relationships towards other countries.

Another narrative focuses on the issue of climate change as a political topic. This narrative underlines a finding by Maibach et al. (2010), which states that the environmental crisis used to be framed as an – in fact – environmental problem. In a recent development, however, it became framed more and more as a political problem.

In one version, this narrative connects the topic of climate change to the role of a nation and its political duties, in the other version the politicization of climate change is put at the center.

CLIMATE CHANGE AS PARTISAN DISTINCTION

Content

This narrative plays an important role among American climate advocates. Its main topic revolves around the traditional disagreement on various issues between the Democratic and the Republican Party. Among other issues, like public health care or stricter gun control laws, climate change is seen as a problem that divides Republicans and Democrats even stronger:

> On the federal level climate change is a totally politicized issue" (I-USA 6: 24); "It is still a very highly politicized in America, and that is a shame because it's bringing a lot of road blocks. (I-USA 2: 10)

In this feud, the Democratic party is established as the political player that takes the threat of climate change serious and is willing to get involved in the fight against it. But even here climate advocates have to deal with disappointment:

> You know I elected Obama thinking that there will be more leadership on this than there has been, and it makes me sad, that, you know… we have a Democrat in the White House, and we should be able to be doing more leadership on this issue. You know, it is sad! I think it is an opportunity lost, and it is… unfortunately it is in a lot of things where people who were most enthusiastic about Obama's leadership were hoping that he would take some more risks. I mean, I will vote for him again, because there is no good alternative, but, anyway… (I-USA 1: 46)

The Republican Party on the other hand makes itself out to be the party that is not gullibly buying into the hype of global warming and thus opposes to take measures right away. In the view of climate advocates this of course is seen as ignoring scientific facts and in the long run putting the lives of others willingly in danger.

> For example, the Republicans who are running for president right now, for the Republican nomination for president, I just can't understand on this point how they can deny climate change when it challenges not their existence but the existence of their children. I mean, even instincts are about protecting your children and your grandchildren and they're flying in the face of that and I think that the Republican party has to re-address their positions on climate because at this point they look like a fool. I mean this is a fact and you can't negotiate with science, and science changes yes, but the scientific consensus for the past ten years has been moving in the direction of climate change is a very real thing and even becoming more and more real. (I-USA 7: 36)

Interviewees put science as the basis for taking measures against climate change. Thus, opposing to or ignoring scientific facts is seen as one of the big obstacles climate advocates have to overcome:

> You know, people are… the whole question has become more similar to a religious belief, either you believe in it or you don't. And science has not a lot of wake with a lot of people. No matter how many charts and data you show them. They either don't understand it or they think, you made it up. That is with certain sections of the population. (I-USA 2: 15)

Depicting climate change as a believe system rather than a result of scientific studies leads to administrative restrictions:

> You know our congress isn't doing really anything productive to respond to climate change. I just read this morning that NOAA, our National Ocean and Atmospheric Agency, was proposing to shift the way it was organized so it would allow for a division of… there would be a national climate service where anyone from farmers to business people could go to find out short term climate predictions and long term climate predictions. They weren't asking for any other money, they just wanted approval to be re-organized. And they got killed in the House of Representatives, because they said it would be a place for propaganda. (I-USA 2: 10)

Similarly, concrete measures that would help vulnerable communities are blocked in political debate. One interviewee explained how she approached a representative of the Federal Emergency Management Agency (FEMA) at a congress:

> I said: I now take the privilege of the chairman of the session to ask a question myself: Could the gentleman of FEMA, please tell me when are they starting to take into account sea level rise in their mappings? And the response was: As soon as congress demands it. That was a totally political answer which means given the national non-consensus about climate change in this country is of course saying we will never get to it in a foreseeable time. (I-USA 6: 24)

The politicization poses a vital challenge to the quest of persuading politicians and citizens to take the threat of climate change serious and to develop measures against it. Thus, the storytellers find themselves opposite an audience which is skeptic of science. Climate advocates accept that the nature of scientific findings is merely absolute. Results of scientific studies do not present an absolute truth but are always subject to debate and change on basis of newly achieved scientific knowledge. This nature of science presents a problem when translating the scientific debate about climate change into the public sphere: "If you say we don't know everything, they could interpret that to mean we don't know anything" (I-USA 2: 24).

Climate advocates draw on personal experience to illustrate the challenge they are facing:

> And science has not a lot of wake with a lot of people. No matter how many charts and data you show them. They either don't understand it or they think, you made it up. [...] You see, my husband is a Ph.D. scientist, and it took quite a while for him to take it seriously. He said, I want to make up my own mind, I don't care if 100 people have said it or 2000 people have said it, I am going to make up my own mind. [...] He does not want anybody to tell him what to think. (I-USA 2: 15–18)

Similarly, local developments are creating an opposition between climate skeptics and climate advocates, as this story told by an American climate advocate shows: a water rescue team proposed building a boathouse at the edge of the jetty in order to quickly reach people in need on the water. However, as sea level is rising and events of heavy precipitation are increasing the jetty is often flooded and inaccessible.

> They ignore [my advice]. They voted for it to get it financed. It is as simple as that. I keep telling them that it is nonsense, but they think: okay, this academic guy, what does he know, what does he understand about our need when we go out there and rescue people?! (I-USA 6: 26)

The Republican Party is perceived as being at the core of science skepticism. Instead of trying to convince their audience of the high credibility of scientific proof of climate change, one way climate advocates suggest to deal with this skepticism is to stay clear of the concept of climate change just to get mitigation and adaptation measures implemented:

> So, we are facing some really big challenges and we are finding, as we move forward that the more we focus on the impacts, whether it is flooding or heat, the more people can connect to it because they experienced it, rather than saying climate change… that seems to get more attraction with our massages and getting people to engage in conversation, but there are still

times where if you say the word climate change… you know. Our governor right now doesn't want to make this a centerpiece at all. But he is very interested in helping communities respond to flooding. So you can engage with him that way, but you can't connect that to climate change. (I-USA 2: 10)

Convincing the audience that climate change exists is neglected in favor of taking concrete actions and to overcome opposition, because

We need communities and municipalities to work together. And under our program there is a start. So, we really promote communities working together. At the watershed program we started these inter-municipal councils, managing the watershed and what flows into the river. (I-USA 55)

Structure

In this narrative the antagonism between hero and villain is set mainly in the opposition of Democrats and Republicans. Even though Democratic leadership has left climate advocates disappointed, the party is still perceived as the better option:

You know I voted Democrats thinking that there will be more leadership on this than there has been, and it makes me sad […]. I mean, there is no good alternative, but, anyway… (I-USA 1: 46)

The general political orientation in the local population opens the possibility of being a hero in this narrative.

You know, you are looking at a fairly Democratic town and certainly here in Little Town and then certainly within the state which is also more Democratic, people will agree or will realize that [increasing flood events] are related to global warming. (I-USA 5: 40)

Thus, climate mitigation and adaptation benefits from the diversion of governmental administration:

You know I live in a village, and that is part of a town, and that is part of a county that is part of the state, so we have these multiple government levels. And on the town and county level there are environmental committees to deal with these things as much as they can. They try to bring people together and look for government support, outside support to solve some of the problems. (I-USA 4: 26)

Governmental organizations are depicted trying to implement measures against political opposition, as the incident about NOAA's attempt to re-organize their administrative structure shows. New structures were needed to offer a climate service to the national public to give them the opportunity of information on long term and short-term climate predictions. Even though this proposal was "killed in the House of Representatives" (I-USA 2: 10) climate advocates see it as an important step for climate education.

Heroes fighting to overcome this politicization of climate change are found less on the federal level but more within local initiatives.

> On the mitigation side a small non-profit is operating in the county; it has gathered representatives from, hm, how many different towns? Like all of this county, and they all applied jointly for a federal grant and they got 800.000 dollars and now they are implementing energy measures with this grant. And basically, it was a grant that required a minimum population size, and so they just said, hey, we just going to do it together. (I-USA 2: 57)

Obviously, this narrative is stronger with American climate advocates, since a consensus among German parties is fairly obvious. However, political tradition also shapes the perception of political climate heroes in the German debate:

> We've always had a quite strong green movement in this state, and today that is even more established with the new state government. In Kleinhausen we had a green party mayor for eight years, so these topics are highly appreciated and dealt with. (I-GER 7: 54)

Lobbyism and shadow politics are blocking climate mitigation and adaptation measures, much to the frustration of climate advocates:

> There is also a very strong political interest, by people who benefit from us not addressing climate change. [...] and they hire very talented people to market their points of views to create this large umbrella of uncertainty. (I-USA 2: 23)

Climate advocates refer to the inability of the main political parties to reach a consensus – i.e. the Republicans to accept the scientific consensus – as the central reason for the impediment of national climate politics. Interviewees refer to congress not doing anything (I-USA 2: 10), not finding a working majority on political bills (I-USA 6: 24) and the House of Representatives blocking climate concerning proposals (I-USA 2: 10). The antagonism between Democrats as heroes and Republicans as villains is made explicit when Republicans are allocated irrationality ("I think that the Republican Party has to re-address their positions on climate because at this point they look like a fool" – I-USA 7: 36) and when political party-orientation is intertwined with geographical locations:

> You are looking at a fairly democratic village. And certainly, here in Little Town and then certainly within the state which is also more democratic, people will agree or will realize that, but if you go further into the rest of the country, all bets are off. (I-USA 5: 40)

The victims of these blockages can be found on national level rather than in vulnerable and existentially threatened countries in the distance:

> On the federal level climate change is a totally politicized issue. On the local level where people get hit by it, it starts to soften at the edges and people start to realize they have to do something whether they attribute that to climate change or not, they simply know they get hit by something. (I-USA 6: 24)

The concern about future generations is also tied to national boundaries, as one interviewee puts it when accusing Republican voters and party members to neglect the issue of climate change even though it challenges the future of their children and grandchildren (I-USA 7: 36).

CLIMATE CHANGE REFLECTING THE ROLE OF A NATION

Content

Besides contextualizing climate change nationally, climate advocates also put political responses to the environmental crisis into a global frame of reference. This narrative introduces other nations and depicts their stand on the topic:

> What is really fascinating to me is to see what other places are thinking about climate change. You know ... Russia is saying, climate change is a big thing for us. Pakistan – they are thinking about climate change. The US are like, no, this is not happening." (I-USA 2: 37)

Politics of countries are compared and some nations are assigned specific functions within the political decision-making process, such as European countries or the European Union in general:

> It is interesting to me to see how quickly Germany can shift when presented with a problem. It's like they say, okay, we know this now, we know about the problem, we now shift, we're going to go this way. That's what I have observed about Germany. It seems like the citizens have really faith and trust in the government to make the right decision. And we don't have that here. We have a lot of skepticism here, people think the government can't make a good decision anyway. (I-USA 2: 47–48)

> In Germany, or Europe, we have to take on the vanguard role. [...] Our technology is highly demanded in the world because we are at the forefront of the technological development. So we need to stay there, and for this taking on the vanguard role is necessary and supported by the citizens because we can afford that. (I-GER 3: 32)

> And I do think that the European Union has to take on a role model function, especially when it comes to climate mitigation. (I-GER 1: 24)

The difficulties of reaching a unified climate treaty are seen as based on nation's different political agendas and opposing interests.

> At every of the global climate summits there were some nations that refused to sign the Kyoto Protocol, for instance the US, Russia, China, and India. These nations have different agendas, the US and Russia as global powers, China as most populous country like India. (I-GER 3: 30)

> I think, recent climate summits have shown how hard it is to reach a consensus between different nations [...] They have all completely different interests, threshold countries, industrial countries, countries that are simply vulnerable to climate change. So, of course they have opposing interests. So, it's really hard for them to come to terms. (I-GER 8: 32)

And in line with the perspective that some nations have to take a role model function or the vanguard role climate advocates expect those countries to work on overcoming opposing interests that block climate policies:

> I really reject this argument, which is also often made by our government: 'We can't do anything unless there has been a European arrangement

beforehand', for instance concerning aviation gasoline; that is just ridiculous, because if they think like that, nothing is ever going to happen. The degree of international interweaving is so exaggerated, just to pretend that the national radius of operation is absolutely small. But it isn't! So, I think that is often just a really convenient excuse. (I-GER 8: 34)

The relation between storyteller and audience as well as their cast is vanished from this narrative. The storyteller is merely made explicit, only in terms of attacking current policies: "I really reject this argument, which is also often made by our government" – I-GER 8: 34; "What is really fascinating to me is to see what other places are thinking about climate change" – I-USA 2: 37. There is no audience addressed directly. The critique of failed and disappointing political action is rather expressed in an abstract, non-personal way.

Structure

Heroes and villains in this narrative are cast from one pool of nations and policy makers. Nations are standing in as pars pro toto for politicians:

> It seems like people are taking this issue very seriously in Germany. Yea, Germany makes major shifts in policy, right? I mean there is this whole green energy policy in Germany and then the recent decision to go away from nuclear power (I-USA 2: 47)

Interviewees emphasize outstanding performances by local politicians or environmental activists. Their actions are seen as a successful example of overcoming national hesitation:

> And you have to see those mayors, like the one we have in [a nearby town], who has won – I think as one of the few communities in Germany – the European Energy Award in Gold. That mayor has been on that energy topic for 25 years, he has done a tremendous job. So, you see: a mayor can achieve a lot if he is really eager. (I-GER 2: 40)

> I met so many people, so many amazing activists, I met so many amazing policy makers. I hung out with the president of – , and I learned a lot from him, and from everyone I was there with, that the solution to climate change isn't going to happen in one day, but you need as many people as possible committing their lives to that process of to getting a fair, ambitious, and binding treaty. (I-USA 7: 30)

Local politicians have to push national decision making by going ahead, because "in general, lower governmental levels can be more efficient and successful" (I-GER 8: 34), even though steps on a smaller scale might lack in global impact, but it is considered an important accomplishment:

> And we will continue that here. We might not save the world with that, but we can contribute our share to climate mitigation. (I-GER 2: 64)

Local initiatives serve as role models, showing how climate change impacts need to be addressed and dealt with:

> And we have been trying to take steps to engage the city to thinking about their long term... hm, I shouldn't say "engage the city", the city is way ahead of us on this. They have tremendous resources there, they already completely understand that it is happening, and they plan for it. (I-USA 2: 33)

Despite disagreement over specific strategies, the fact that a local government is taking action beyond national standards is highly valued:

> They would be considered as the national and global leaders in responding both in mitigation and adaptation, but some of the way they might choose to respond hits back to the solution. (I-USA 2 33)

As noted above, the villain in this narrative is also drawn from a pool of nations, with politicians not making an effort to minimize the impacts of climate change.

> I went into the conference hoping that we were going to get a fair ambitious and binding climate treaty, and I mean fair for... emphasis on the word fair, because some of them,... on that level regarding climate change has been dictated by the developed world, dictated by United States, dictated by the EU and by China and India. [...] The way the UN is structured unfortunately the developed world is favored from a policy level and obviously favored from an economic level. (I-USA 7: 30)

Among these nations, the USA are outstanding out from the pool of villains. Climate advocates, both from Germany and the USA agree that U.S. American politicians fail in addressing climate politics.

I wish, as a citizen that we were much more taking a leadership position, and we are not. And there is a lot of political reasons for that and a lot of global economic crisis reasons for that and a lot of reasons and a lot of excuses. (I-USA 2: 46)

The U.S. is an example: the U.S. can pretty much buy their way out of having to directly address climate change. (I-USA 7: 30)

The leadership role, which in other political areas is often assigned to the USA, and their performance within global environmental politics is criticized on the one hand with a national frame of reference:

We have a Democrat in the White House, and we should be able to be doing more leadership on this issue. You know, it is sad! I think it is an opportunity lost, and it is… sad. (I-USA 1: 46)

Well, we are not leaders. America is very America first, even though we are missing a huge economic opportunity to be America first. We could be setting ourselves up be a leader in renewable energy and adaptation strategies, and we are not taking advantage of the biggest opportunity that we have. (I-USA 2: 35)

Critique is also formulated against the backdrop of other nation's politics and their role in the debate:

It is pretty clear that we have a problem, and it is also clear that we only gonna slow down the problem if every nation realizes it. And when it comes to that, the USA are as bad as others, nations like Russia or China. […] And if governments do not prioritize the issue of climate change, […] Then we are not going to get any results. And personally, I don't see anything happening. (I-GER 7: 56)

The role of the villain is directly linked to the role of the victim, showing how the lack of political action and the refusal to take responsibility in the environmental crisis of some of the industrialized countries is hurting those who are most vulnerable to the impacts of climate change. However, here political defenselessness defines the status of being a victim.

I went into the conference hoping that we were going to get a fair ambitious and binding climate treaty, and I mean fair for… emphasis on the word fair, because some of them,… on that level regarding climate change has been dictated by the developed world, dictated by United States, dictated by the EU and by China and India. And you have all these countries that are part of the G77 and all the weak developed countries kind of squabble in because they have no chance under the regimes that are being enacted by the developed world. The way the UN is structured unfortunately the developed world is favored from a policy level and obviously favored from an economic level. […] But a nation like Tobalu or pretty much any small nation, nations like Tanzania don't have the resources to even feed their own people let alone to address climate change. (I-USA 7: 30)

As in other narratives, climate advocates establish the contrast between villain and victim not only with the opposition of powerful – powerless, but they emphasize how the developing world is mainly suffering from an impact, that the developing world caused:

How did we manage our industrial development? Over the last century? And now we are permitting those developing countries the use of hard coal and brown coal. And nuclear energy. […]And do not show them an alternative way? So, we basically developed our societies on their backs (I-GER 7: 58);

I am talking about those [countries] that do not have any economic stakes in that, but have to deal with the impacts, like droughts. (I-GER 8: 36)

THE WORLD IS (NOT) ENOUGH: ENVIRONMENTAL CONCERNS AS A TOPIC OF MORALITY

This narrative centers on the values of environment itself. This shows the meaning interviewees assign to a specific place. It intertwines the sense that the earth is unique and precious and has to be protected and the relationship between human beings and the natural environment. This way, the value of nature is directly linked to social culture.

Content

This narrative centers on the values of environment itself. This shows the meaning interviewees assign to a specific place. It intertwines the sense that the earth is unique and precious and has to be protected and the relationship between human beings and the natural environment. This way, the value of nature is directly linked to social culture.

There are basically three key-aspects of the main topic in this narrative: the historical influence of a specific place, the meaning of a place to the people experiencing it every day, and the relationships between human beings and the environment.

The region of the US American case is an iconic part of the landscape and is recognized for its uniqueness:

> You know, there is a fair number of people who are very passionate about the river because of all the work that has been done and because it is just an iconic river. (I-USA 1: 10).

Even though the grandness of the landscape is note cited directly as a reason for pursuing climate politics it underpins the motivation and value system of climate advocates. The value of nature itself is found in the river's history, influencing cultural and social developments. Interviewees characterize the river and its valley as a special place worth protecting especially because of its unique nature and its unique meaning, not only for those people living there but also for those who come to visit.

> Every night is a different back drop, you have a different cloud pattern, every night you go out there, some nights it is incredible, some nights it is blah, but it is always different. You never see, unless it is crystal clear, you never see the same backdrop. So, every night it is dramatic. […]That is great around here, not only for painting. (I-USA 3: 17)

The river's entire ecosystem needs protection. Here – in contrast to the narrative of economic consequences – not as means to an (economic) end, but as a value in itself. One of the interviewees who works for a river protection NGO and who is also involved as an environmental activist puts it as follows:

> People usually understand it, especially when you're talking about fish, people start to understand it. People like fish. It is kind of funny, but people are actually having an emotional reaction to the fact that fish species' life chances are threatened by this issue. It is funny because you could say that their lives are being threatened as well, but they are having more of a reaction to fish. I think people value their diversity, I forgot what the term is, it's an economic term, something like non-use economic value [non-use value] and it's weird, because people are not eating the fish, they are not even seeing them, but they like the idea that they are there. For the same reason people

like polar bears. I mean, sure, polar bears are cute, but… people like all these different types, people just like biodiversity, I don't know how else to explain it. And, I mean, a sturgeon is not cute and cuddly but there is something special about it, you know. And to know that something has always been there, and that it is not okay if it's gone. Especially something like the Atlantic sturgeon which has lived in the system for millions of years. I hold the sturgeon thing as a symbol to my heart for the system that had had this huge impact on me. I think they are a very important species, I've seen one once. I don't often, but I value them, even though they benefit me in no way. It is just an emotional capacity, I just value their existence. (I-USA 7: 40)

This quote also marks the entrance to the topic of the relationship between humanity and nature, fleshed out in the description of the relationship between people and the river. The river is part of the lives of those people living in the valley ("You can't live in […] without having some interaction with the river." – I-USA 7: 24) and climate advocates describe their work on the issue of climate change with reference to the river, thus turning climate work into concrete, local action that is done for the river, as opposed to climate work as abstract politics.

Our job is to protect and revitalize the river, we look at the river as a whole ecosystem, the river, the land it crosses and its surroundings, the watershed, and we are looking at all the different influences on the river that are affecting it today. So, we deal with everything from cleaning up the water quality of the river, try to make it safe for swimming, restoring the fisheries – some of them are very depleted, some of them are robust, and make sure, that those species stay robust, and those that are endangered do come back, we are trying to provide access to the river, so that the public can enjoy the river, for swimming, hiking, boating, picnicking, glazing at the sunset, painting. Big culture. Protecting the aquatic habitats and the land habitats of the valley, so that we have a great diverse ecosystem, that is resilient in context of change. So, as part of that, climate change is the number one factor. It affects how we use the river, how the river sustains us, life itself. It is a key-aspect of that (I-USA 2: 4)

The value of nature is set in relation to the social culture, seeking a balance between both spheres as a "mutually beneficial environment" (I-USA 7: 26). People need to take care of their environment so they will be able to enjoy it in the future, as the quote above shows (I-USA 2: 4: "So that the public can enjoy the river, for swimming, hiking, boating, picnicking, glazing at the sunset, painting. Big culture"). This relationship is proffered as reason to get involved in climate politics as climate change impacts are influencing social culture:

> I talked to some anglers today, and they told me that the fish are suffering from a lack of oxygen when the rivers are getting warmer. So, I learned that our rivers are on average three degrees warmer now in the month of May than they were in the past years. (I-GER 4: 34)

Climate advocates emphasize the reciprocity of the relationship between human culture and nature to bring the value of nature to people's minds:

> It used to be fact that the river was so toxic that you wouldn't want to get near it, but in recent years you can see the return to the river, and how it gets cleaned up, and I like to think that part of my position is educating others on the fact that they can interact with the river and the river whether you know it or not provides for you in some way. (I-USA 7: 24).

In one interviewee's outstanding claim nature and human lives are even seen as of equal value:

> And I don't value – and maybe that is a difference to other people – I don't value human life over other life forms, and I know that, or, well, definitely, that's, what most of the world does, but I really don't value my own life over the life of others... (I-USA 7: 32)

Storytellers are aware of their audience and that the topic of this narrative has to be told carefully to persuade.

> It's been the challenge of my life so far. One method I use is actually story telling. I mean, I love policy work, I enjoy policy, but what is lacking in the environmental movement is the whole of emotion, I tell the story about my time in Copenhagen and then about my time in Tanzania, and emotions are more powerful than any policy documents, emotions have the capacity to drive people's personal action, and I find that when you engage with people on the personal level, when you tell them your story, they have a reaction. [...] I talked to a lot of people in the course of a few months I talked to a lot of people about climate change specifically, and you know for people who are either deniers or just don't care that much, the personal story always means something. (I-USA 7: 34)

Making the story personal on behalf of the storyteller also leads to addressing the audience on a personal, emotional level.

> [...] Everyone has a spot I think, they've been to, they can think about like 'I just felt free there' [...] I mean, it's been psychological to me, when I speak in public about my experience with climate change and witnessing it people are moved you know? By that! They want to know what they can do. (I-USA 7: 36)

The meaning of a specific place to an audience is understood as a way of reaching people's emotional relationship to that place and the historical development of a place encourages this way of addressing:

> People stopped looking at it as an industrial waterway and started thinking of it as a recreational source, which is tremendous! It's wonderful. And by going there you found people instead of looking to live nowhere near the river because it was horrible, it smelled bad and it was ugly, they started to think about it as a very beautiful place to live and wanted to get near and closer to it. (I-USA 5: 20)

In this approach, the storyteller is extremely present in this narrative, as she draws on personal experience to tell the story of a place's beauty and meaning and the value of nature itself:

> You know, I grew up in this valley, I grew up in [...], which is on the river, and the river has always been part of my life. I've never lived outside the valley, apart from when I went to school, and it is a really big part of my life, just day to day, it was around me, it's what inspires me. (I-USA 7: 18)

Structure

The heroes in this narrative are those actors that recognize the beauty of landscape and nature as a value in itself and set out to protect it. This becomes part of the history of environmental protection:

> [Thomas Cole] fell in love with this area so much, so he eventually became one of our first environmentalists later on, because probably from the early 1800s to the 1900s they completely stripped this whole mountain of timber. If you look at black and white photos of this area in the 1890s to 1900 it looked like a wasteland, it looked like an atomic bomb went off here [...]

And this and many other tanneries throughout the Catskills were responsible for stripping this whole area and then the railroads came through. That's another thing Thomas Cole spoke out against, these railroads. He just saw no beauty in it. (I-USA 3: 4)

Environmental activists, NGOs, and citizens that get involved in local environmental tasks are set in relation to specific local environmental problems, like the pollution of a watershed, etc.:

> You have some active people, volunteers, that sort of try to clean up the Sparkill, and when they do that they become aware of flooding etc. (I-USA 6: 30)

> There are organizations involved in these things a great deal. And I am part of one of such a group, which just has recently started to monitor the creek, our local creek that empties into the river that has these huge pollution problems. But, yea, there are lots of organizations. I think communities and small towns and villages have now environmental committees, they are not as effective as they need to be but at least, they exist. (I-USA 4: 26)

Being in contact with nature becomes a characteristic of heroes in this story, for "they just see that some species are disappearing and that new species are showing up" (I-GER 8: 6). A development that some interviewees see spreading among the society:

> That people are more aware of their environment and that they appreciate it more. And then you have the whole issue of consuming local products. (I-GER 1: 48)

Contrary to that the character of the villain is assigned an abusive posture towards nature or at least a lack of appreciation:

> I just love living things and I think that unfortunately over time humans have lost sight of our role on the planet, which is not... I mean our role is not to be dominant over other species. Right now, that is totally forgotten. People abuse the environment because they think they are better than it, but it is not a question of being better or worse, the question is about cohabitating. (I-USA 7: 26)

> But I think people living in the urban areas or white-collar workers are not that aware of this problem yet. (I-GER 8: 6)

The villain's ignorance of nature is blamed for the situation the victim finds itself in. Here, the victim clearly is the endangered environment with its habitants. These losses are described with specific local reference:

> It is also changing the salinity content of the river, the river is in a constant give and take between the ocean being salty and freshwater from the mountains and if the salinity balance gets changed you know the biological makeup of the river is going to change and I am sure it already has and I am sure there are species that no longer prefer to live in the river because of that, and I think that is inexcusable. (I-USA 7: 38)

> Here in Little Town we just recently discovered that this one creek was just this beautiful creek which is in many of my paintings, that is intensely polluted. In fact, it is one of the most polluted sites in the state. And that just shocks us. And that is because of a lot of thing, but also because of an increase in the water, additional rainfalls and the fact, that we get these new sorts of storm surges, and higher tides, that polluted water has become very invasive. (I-USA 4: 12)

The relationship between people and nature is once more strongly emphasized as a reciprocal one and at the same time nature is portrayed as something in need of the hero's protection. Nature suffers from the villain's intrusion:

> We all live in this one world, this one earth, we all are dealing with the sun and freak weather… it has to be clear to everyone that something in the atmosphere has changed and that these changes are anthropogenic. (I-GER 7: 58)

> And almost all disasters are caused by mankind, at least partially. No matter if it is fire, wildland fires, or floods. I think especially floods are directly related to anthropogenic activities. (I-GER 4: 65)

Moreover, climate advocates insist that damaging the environment will come back to those who caused these damages, thus the narrative structure causally interlinks all of the figures:

> The river is an estuary, […] and so sea-level rise as part of climate change is already affecting us here, it has gone up 6 inches from when I was a teenager, and it is expected to go anywhere from two to five or six feed in the next 100 years due to a increasing sea level. And that is going to affect the ecosystem, it's going to affect human communities, it's going to affect our ability to live in the valley. (I-USA 2: 4)

About the "Global" in "Global Warming": Solidarity and Responsibility as Motivators

The core in this story can be summarized with the idea of "think globally, act locally". Climate change is not a problem, that can be tackled and solved within national boundaries. It needs a coordinated international effort and climate advocates are very aware of this; at the same time, national governments need to acknowledge their role and take on responsibility. The following narrative tries to bridge the gap between global responsibility and local action, between international agreements and personal solidarity. On an action-based-level: they need to overcome the time-spatio distance of climate change to inspire people to get involved on a local level. The narrative consists of two main topics that are intertwined: first, advocates invoke the idea of global solidarity and empathy with those who suffer in distant regions from the impacts of climate change. Second, they need to find a way to connect this distant suffering to personal and local responsibility. But these two topics have different origins in the story. The notion of solidarity derives from the idea of humanity, not from a sense of guilt. People should get involved in climate mitigation because other human beings are suffering from the impacts:

> And there are so many realms, those famines in some parts of Africa, weird weather incidents somewhere else... shouldn't there be some kind of initiatives, people should unite and realize that this affects us as human beings, not as an industrialized nation, but as human beings. My existence, my life is going to be difficult in the future. I don't understand why people don't seem to realize how hard it's going to be! (I-GER 7: 58)

> But I do believe that these people have the right to continue to do this and to live and their existence is just being challenged directly by burning fossil fuels. I take personal offense to that. I think that is why people should care. (I-USA 7: 36)

However, climate advocates are aware of the problem of distant suffering and how this affects people's willingness to help:

> I mean, on an individual level people can be very helpful. I have relatives in Haiti, when there was this earthquake in Haiti, everyone donated something. I don't know anybody who didn't, donate money or whatever. But that got such tremendous press, and other things, like the monsoon in Pakistan, yes, they got press, but not the same level like the disaster in Haiti. I am sure a lot of people weren't even aware that they happened. (I-USA 2: 35)

Structures that come with distant suffering[7] play an important part in this narrative as they are seen as an obstacle for a lot of people that might otherwise be willing to help. The problem of the time- spatio-distant character of climate change comes once more into play:

> You know, I think every human being wants to help in some way. If you have a situation like that at home it's easier to help, because it is tangible. In cases like Haiti or Somalia, there is so much information and are so many organizations, where money goes to waste in the administration, there are so many opposing interests... Personally, I think that makes it hard to help there... (I-GER 7: 48)

They also recognize a struggle for attention each crisis has to win. A constant stream of reports about people's suffering in TV news, the internet – including social networks – and newspapers is wearing thin the audience's attention:

> And I think to a certain extent, disaster after disaster after disaster people just stop feeling as powerfully, like, okay, now, this is going to be every day. I can't deal with this every day. (I-USA 2: 37)

> A drought in Africa or an erosion in Pakistan on the other hand, I fear people don't care about that. Or only if they have some kind of connection to the country, or know people who are affected. Personally, I feel that people need to be affected on a personal level and that reports people only get blunted by TV or radio news. They don't think "oh, right, that is climate change, we have to do something about it" or something like that. (I-GER 8: 18)

Audience and storyteller are depicted as being in the same situation:

> We all live in this one world, this one earth; we all have to deal with the same sun or some kind of freak weather. [...] I do not understand how people don't see that everything is going to be harder. (I-GER 7: 58).

Structure

Being a hero in this narrative implies to address climate change for the sake of those who are suffering. But these heroes are doing it not due to guilt, but because they feel a moral obligation towards fellow human beings:

In some parts of society, take the churches for example, they always point out that we have a moral responsibility, and people who are involved in church activities they might really feel that responsibility. And the environmental NGO's also hint at that. (I-GER 8: 44)

Since the distance between those who can help and those in need of that help is an obstacle, heroes are assigned a power to overcome this obstacle by drawing from personal experience. This way, the gap between those in need and potential helpers (heroes) can be closed:

We look at flood events differently. With the perspective "Can we help there?" For instance, the Elbe flooding in 2000 made us think about what we can do to help. [...] And we experience that here in our company: Employees ask to be released from work because they were asked to help in Dresden when they had this big flood. Of course, on the one hand it doesn't help the company, but it is important for the case. And I think the common interest has to be rated higher in that case than my company's interest. (I-GER 7: 46)

Climate mitigating action that is taken on a local level is put into global context, emphasizing how local initiatives can make a difference on a global scale. With this, the vulnerability of geographically distant places is brought into the every-day-life experience of the hero:

So, this was a regional planning process, and I know that there are a lot of regional global climate change processes, because if you try to address the issue on global level, there are a lot of players doing that and doing that well, with their conventions and efforts that are underway country to country working on various protocols to reduce emissions. The thought on this was, our piece of what we could do to deal with global climate change is make the region a better more adapted place to live. And that will help in the overall global theme. Because everything we do to reduce climate change impacts and to reduce the impacts on the river, because we are a big consumer of resources and a wealthy place, there is a lot of reasons why we should be able to do it well, because we have a lot of resources to think it through. And that is replicable in other places. (I-USA 1: 38)

Personal and direct encounters play a key role in local efforts. Activists stress the importance of being approachable and visible in the community. Every-day work becomes a conversation starter, which also helps de-politicizing the subject of environmental conservation:

Yes, absolutely. It is a huge part of the reason, why the boat is so attractive as an enforcement tool because it is a conversation starter, so, when people see us out on the water, whether they are boaters, or kayakers, or just interested citizens, they want to talk to us and oftentimes when we're on patrol we get as close as we can to land to have a conversation with people because they want to know what we're doing, what we are working on and how they can get involved. (I-USA 7: 12)

The experience of helping others is also made on a local level, thus serving as a role model for helping others where the geographical distance and the distance between experiences of everyday life is even bigger:

I am not that affected, I am concerned because I am a citizen and this is my community and it concerns me, and as a community we really try with our own money and grants to protect residents. (I-USA 4: 14)

The suffering of the victims in this narrative is told as a personal, intimate story. In the following quote this is exceptionally done by retelling a personal meeting and actually using the words of the victim:

But then he mentioned the Copenhagen conference without me ever mentioning it, and he said, you know, I see these people in Copenhagen and they don't care about climate, they don't care about our way of life, they don't care about me and my children, they are in it for the money, [...] You need to go back to the U.S. and you need to tell people that we are dying here. I mean, this is life or death for us. We have no options here. (I-USA 7: 36)

The villain in this narrative is characterized by ignorance of the suffering and negligence of the victims, thus adding to their suffering actively by, i.e. polluting or preserving egoistic interests, but by way of being inactive in the face of disaster:

People aren't just that altruistic in our society. Own interests always come first. (I-GER 8: 48)

Historic Responsibility

Content

Climate change in this narrative is viewed as a consequence of the way industrialized countries developed their wealth and standard of living on the expense of other countries. That indicates that industrialized countries have a responsibility to provide aid to those countries that are suffering from the impacts of climate change. This help is not seen as actually providing financial and development aid, but is contextualized as a reason for industrialized wealthy countries to get involved in climate mitigation:

> Countries like the USA or European countries, which are emitting CO_2 beyond the bearable amount for decades; those countries have the responsibility to counteract und to support the most vulnerable country. That's a given. We have this huge obligation to provide. We caused climate change, and those threshold countries are on the best way to catch up with us. But we are responsible for the larger parts of this development that lead to global warming. (I-GER 8: 40)

> It's us, these highly industrialized countries of the West, the U.S., European countries, Australia. Brazil is catching up, but in a historical perspective we are responsible for those CO_2 emissions over the last decades that caused climate change. (I-GER 8: 42)

The main topic in this narrative resembles the profligacy story in Ney and Thompson's (2000: 73) study, where the lifestyle of over-consumption in industrialized countries is made responsible for today's global warming. The following quote form a German interviewee brings in the concept of a community of responsibility which has to be established among wealthy and well-developed countries:

> So, in our processes there was always a lot of talk about a community of responsibility. And I feel that captures it quite well. No matter if you look at the problem from a regional, a national, or an international perspective: in general, it is always some kind of a community of responsibility. And, the question of blame, well, that's always a popular game, that doesn't lead anywhere. (I-GER 5: 44)

At the same time, the problem of climate change is connected to the local challenges on site, creating a frame of reference for local experience. The historical development thus cannot only be put in terms of economic progress but in terms of geographical and cultural history:

> But right now, I think specifically the problem with climate change here on the edge of the river is, what is it that you can really do? How do you hold

back the ocean and that is the question that mankind has been dealing with many years. (I-USA 5: 46)

That is what happens. We are not a nation that started out with dikes, we are not the Netherlands that is not build into our consciousness that we are going to build these barriers and find a way to live and to adjust. Historically we are not a nation that has done this, so we don't have the infrastructure we don't have the history to fighting back the sea. (I-USA 4: 16)

The audience is addressed by providing concrete suggestions for contributing to climate mitigation:

So we are trying to show again and again where every individual can cut back on CO_2; mobility, nutrition, consumerism in general. And of course heating and living. On national average every citizen emits 11 tons of CO_2 per annum, and it is really hard to cut back to a bearable amount. But in our project we do have families who already reached a level of six or seven tons. But, of course those are interested in these issues anyway... (I-GER 8: 10)

In education on climate mitigation this kind of concrete suggestions is seen as very helpful and is in line with findings from studies on environmental communication (Aronson 2008).

If you offer people specific advice on what to do and about funding possibilities, so this practical use... I think that involves everyone's realm of experience and people respond to that. That is more use than blank theory. (I-GER 8: 30)

However, in the overall topic climate advocates see the problem that blaming people may not have the desired effect for the audience will feel (wrongfully) accused and will get defensive:

The problem is, that all these things, when you talk about them, they involve guilt. Because everybody has their own air condition, their own car – including myself – and while some people are more conscious about it than others it makes you feel guilty when you are being told you are wasting energy. And that is where the personal issue comes in. People get defensive right away. (I-USA 6: 28)

Structure

Action has to be taken by industrialized countries against the backdrop of this responsibility. Local initiatives and environmental groups are seen as actors in this field that take on this challenge.

> They try to bring people together and look for government support, outside support to solve some of the problems. People are aware of it and there are organizations that participate. (USA 4: 26)

> But in our project, we do have families who already reached a level of six or seven tons. But, of course those are interested in these issues anyway… (I-GER 8: 10)

Interest in environmental issues and in getting involved in climate mitigation is also connected to personal experience and a sense of responsibility:

> Well, I think, my concerns, everything relates back to my love for the river and the valley, I think that's is the lodging point for my inspiration, but I mean it definitely goes beyond that, I mean, I have a general… I mean, I haven't been all over the world, but I have travelled the world a bit, and I was actually living in Tanzania during my junior year, I certainly haven't been all over the world, but I guess I just have a general love for the planet. (I-USA 7: 25–26)

The victims in this narrative are clearly those who are innocently suffering from the impacts of climate change, caused by the industrialized world. The reason to get active and protect those who suffer lies in the wrongfulness and injustice of the situation:

> But you have to consider the fact that people live where they live for a reason and we have no right to say 'you can't live here anymore'. (I-USA 7: 32)

In opposition to this, the villain is accused of ignoring her responsibility and not getting action on the way:

> And if none of the governments understand that every missed reduction goal goes towards making it harder to live on this earth… (I-GER 7: 58)

It is also stressed in the characterization of the villain that the responsibilities towards vulnerable countries are based on historical failure of industrialized countries.

But the reason why we should take an interest in climate change is that we are accelerating it through burning fossil fuels, and it is creating an environment, that is unlivable for things that would otherwise thrive in a stable climate. […] But we are putting so much of that into the atmosphere that the earth just can't keep up. And I think it is a fallacy that humans feel entitled to destabilizing an otherwise stable climate. (I-USA 7: 32)

This narrative resembles what Liverman calls "common but differentiated responsibility" (Liverman 2009: 288) which refers to conflicts over assigning responsibility for global warming, centering on north-south-relations and the amount of CO_2 emissions, past and future.

Notes

1. To ensure anonymity, references to specific places, cities, persons, etc, that were made in the interviews, have been altered accordingly; these changes do not influence the content of the interviews and the quotes, to which I refer throughput the study.
2. German quotes translated into English.
3. This concern can also be found in the social science literature on climate change, where scholars emphasize that fear messages often might lead to unintended and unwanted consequences (Moser and Walser 2008; Ereaut and Segnit 2006; Aronson 2008).
4. Just as discussed in the state of research section in this book, this interviewee is aware of the dangers of fear messages.
5. The interviewee is here referring to the devastating damages hurricane Katrina caused in the city of New Orleans in 2005.
6. KfW = Kreditanstalt für Wiederaufbau (German government-owned development bank, based in Frankfurt, Germany).
7. The problem of distant suffering and how people are dealing with this is – among others – prominently considered in Luc Boltanski's "Distant suffering. Morality, media, and politics" (1999).

References

Aronson, E. (2008). Fear, Denial, and Sensible Action in the Face of Disasters. *Social Research, 75*(3), 855–872.

Ereaut, G., & Segnit, N. (2006). *Warm Words: How Are We Telling the Climate Story and Can We Tell It Better?* London. Retrieved from Institute for Public Policy Research website: http://www.ippr.org/images/media/files/publication/2011/05/warm_words_1529.pdf

Geertz, C. (1973). Ch. 1: Thick Description: Toward an Interpretive Theory of Culture. In C. Geertz (Ed.), *The Interpretation of Cultures. Selected Essays* (pp. 3–30). New York: Basic Books.

Liverman, D. M. (2009). Conventions of Climate Change: Constructions of Danger and the Dispossession of the Atmosphere. *Journal of Historical Geography, 35*(2), 279–296.

Maibach, E. W., Nisbet, M. C., Baldwin, P., Akerlof, K., & Diao, G. (2010). Reframing Climate Change as a Public Health Issue: An Exploratory Study of Public Reactions. *BMC Public Health, 10*(1), 299. https://doi.org/10.1186/1471-2458-10-299.

Ney, S., Thompson, M. (2000): Cultural Discourses in the Global Climate Change Debate. In: Eberhard Jochem, Jayant A. Sathaye und Daniel Bouille Society, Behaviour, and Climate Change Mitigation. Dordrecht: Springer, 65–92. http://link.springer.com/book/10.1007/0-306-48160-X/page/1

Renn, O. (2011). The Social Amplification/Attenuation of Risk Framework. Application to Climate Change. *Wiley Interdisciplinary Reviews: Climate Change, 2*(2), 154–169. https://doi.org/10.1002/wcc.99.

The Canada Institute of the Woodrow, Moser, S. C., & Walser, M. (2008). Communicating Climate Change Motivating Citizen Action. In C. J. Cleveland (Ed.), *Encyclopedia of Earth*. Washington, DC: Environmental Information Coalition, National Council for Science and the Environment.

CHAPTER 5

Conclusions: Pitfalls and the Power of Narratives

Abstract This chapter discusses selected aspects derived from the presentation of findings in relation to the state of research in climate change communication studies. Against the backdrop of cultural narrative analysis this chapter traces recurring and partly contradicting themes and motifs in climate advocates' narratives. Drawing on these analyses the chapter carves out some suggestions for climate change communication, not in terms of manipulating the public debate, but in terms of considering the power of cultural patterns that lie beneath social narratives.

Keywords Narrative analysis • Climate change communication • Climate policy • Partisanship • Narrative characters

In the following chapter, I will discuss selected aspects of the above described narratives. I will first lay out how those themes presented in the chapter concerning the state of research are recurring in the findings of this study and where differences lie. Secondly, I will show how the relationship between storyteller and audience influences the frame in which the story is presented. Finally, following Lamont's concept of boundary work of social groups (Lamont, Fournier 1992; Lamont 2000) I will show how the characters separate themselves from the other character groups within a narrative.

The narratives described above show how a phenomenon based in the sphere of natural sciences becomes storied and what Smith calls "a meaningful social fact" (Smith 2012: 745). This is how climate change becomes part of the social world by mimicking those pre-modern structured myths and narratives our societies are still holding onto.

Recurring Motifs and Themes

The literature overview in Chap. 2 revealed various motifs and themes in climate change communication. Some of these occur again in the findings I have presented above.

The narrative of positive economic consequences can be found essentially in Liverman's (2009) description of the story about climate change as an investment opportunity. Even though the narrative in the present study entails various economic options, like becoming "a leader in renewable energy" (I-USA 2: 35) and other economic benefits, whereas Liverman focuses on carbon-trading as a market solution to climate change, both narratives introduce the environmental threat into the realm of economic reason. This once more underlines how climate change becomes part of the social sphere (see here also Renn 2011: 164).

The economic perspective plays a significant role in the debate about climate change, as is shown by the two economic narratives in this study as well as in other analysis of climate change communication, such as the studies by Ney and Thompson (2000) and Verweij et al. (2006). These studies show how climate mitigation activities become embedded into economic reason. The story of price resembles the story of climate change as an investment opportunity and the narratives of economic consequences. In both studies, the critique on the wasteful lifestyle in high-income, industrialized countries is prominently represented as the story of profligacy and can be found in this study on narratives as well. Here, however, the stories told by climate advocates are further differentiated: storytellers invoke a responsibility towards nature itself, a responsibility towards suffering humans in highly affected parts of the world out of basic human solidarity as well as specific historical guilt (benefitting from the industrial revolution which contributed highly to the environmental crisis). This further detailing enhances our understanding of the motivation behind each narrative. Uncovering various narrative elements, the cultural structural analysis of narratives also allows adding an additional layer to the cultural theoretical analysis of climate change stories as presented by Ney and

Thompson (2000). Various studies (Smith 2012; Jones 2010) describe characters as heroes, villains, and victims. However, they often lack extending their analysis beyond a pure description of those characters and to relate the characters within the story and the surrounding cast (storyteller and audience) to the development and potential of the story.

Arguments made within each narrative however do not entail an explicit opposition of the various realms. For example, the economic narrative does not present itself as a distinction to environmental or social concern; these realms simply are not part of the economic narrative. In her study on Norwegian stories about climate change, Norgaard (2011) identifies two stories that are made to legitimize Norwegian climate policy: "Norway is a little land" – minimizing Norway's contribution to climate change – and "We have suffered" – a story told to show how Norway just fairly recently became a wealthier country due to its petrol industry. With this, the author inherently puts economic development opposite of climate mitigation activities, thus strengthening an alleged and common premise of an opposition between these two realms, and – more importantly – an opposition triggered by the occurrence of climate change. Despite her cultural perspective, this opposition prevents the inclusion of climate change into the social world as a meaningful social fact. Climate change thus remains an outside disturbance to social life.

Topics as Result of the Relationship Between Storyteller and Audience

Narratives consist of the events and characters occurring within the story (hero, villain, victim, main topic) and the cast surrounding the narrative. The latter is what Smith calls the meta-discourse over a narrative (Smith 2012: 758). Every narrative has to be told by a storyteller. According to science communication literature the teller of a story needs to know her audience if that story aims at convincing an audience of a specific task. Hart and Nisbet (2011) insist that communication efforts need to take into account the audience's predisposition on an issue. With regard to climate change communication Nisbet points out "to break through the communication barriers of human nature, partisan identity, and media fragmentation, messages need to be tailored to a specific medium and audience, using carefully researched metaphors, allusions, and examples that trigger a new way of thinking about the personal relevance of climate change" (Nisbet 2009: 15). This can be found in the described narratives, where climate advocates built narratives around various main topics, depending on the "imagined" audience. Climate advocates fit their arguments to topics they imagine their audience will listen to. One striking example for this is the narrative of economy:

> I **always insist that whatever I do, I put a dollar sign on it.** So, when I say what is the probability of such and such to occur, I say well it will cost the village a million dollars per event – or whatever. (I-USA 6: 28)

> I think the only way we can get through the challenge of the politicization of climate change is if we can show people a direct benefit, […] because that is what people respond to, esp. with the economy the way it is. I think, that's the best way, to try to get our message across. (I-USA 2: 40)

> Well, if you really think about it, it is a win-win-situation. If I approach the issue of energy saving, I firstly save money and secondly I get a climate mitigating effect. (I-GER 2: 18)

Communication research literature suggests various frames for climate change communication. For example, a connection between public health and the impacts of climate change is considered successful as it arouses hopeful emotions and thus a positive response (Maibach et al. 2010). Myers et al. argue, that "research on a public health frame, for example, suggests that when climate change is introduced as a human health issue, a broad cross-section of audiences – even segments otherwise skeptical of climate science – find the information to be compelling and useful." (Myers et al. 2012: 1107). The presented data in this study hint at public health to be an issue. However, there is no explicit connection made between climate change and public health, but climate advocates put it in context with a broader environmental issue, such as clean air and a clean river:

> So, we deal with everything from cleaning up the water quality of the river, try to make it safe for swimming, restoring the fisheries. (I-USA 2: 4)

> Somewhat related is this creek that we have running through the village, you saw. Now, sewage will end up finding its way into that creek […] we have a little park down there in the area by the Troy Bridge and a playground. What I don't understand is, once that had been flooded with questionable water what is the proper time frame to say people shouldn't go into that park with

their young kids playing on the swings and what not. [...] so, that's general public issues that we will need to figure out how to deal with in terms of alerting the public when something like that happens (I-USA 5: 25)

The last quote shows how the impacts of climate change touch on public health issues. However, it is not presented as a main topic by the interviewed climate advocates.

BOUNDARIES BETWEEN CHARACTERS

Climate advocates get involved in boundary work,[1] establishing two groups with clearly differing values and goals in the narrative of climate change as a partisan issue. Climate change becomes another topic in which the American political elite is deeply divided, with Democrats blaming Republicans for blocking important decisions in climate politics (as most climate agent represent themselves as Democrats: "You know, you are looking at a fairly Democratic village and certainly here in Little Town and then certainly within the state which is also by nature being a Northeast state and more Democratic will agree or will realize that [increasing flood events] are related to global warming." – I-USA 5: 40; "It is still a very highly politicized issue in America, and that is a shame because it's bringing a lot of road blocks." – I-USA 2: 10). The strong divide between supporters of the Republican and Democratic party is backed by findings from survey reports. Engels et al. report that "for the U.S., social scientists have predicted a growing divide between liberals and conservatives rather than the emergence of a social consensus. [...] The probability of holding skeptical views on climate change is significantly higher among white male respondents who identify with the conservative party than among any other group" (Engels et al. 2013: 1019).

Climate advocates do not try to achieve a new status of this relationship but try to circumvent the differences between heroes and villains by excluding the object of struggle (i.e. climate change). This strategy does not allow for a long-time solution to climate change, since it does not cover all the aspects this ecological crisis contains, as one interviewee states:

> But there are still times where if you say the word climate change... you know. Our governor right now doesn't want to make this a centerpiece at all. But he is very interested in helping communities respond to flooding. So, you can engage with him that way, but you can't connect that to climate change. (I-USA 2: 10)

Common ground is not to be found in this narrative, when climate change becomes characterized not in terms of a rationally discussed topic within political decision making but becomes subject to negotiations of fundamental beliefs:

> You know, people are... the whole question has become more similar to a religious belief, either you believe in it or you don't. And science has not a lot of wake with a lot of people. No matter how many charts and data you show them. They either don't understand it or they think, you made it up. That is with certain sections of the population. (I-SA 2: 15)

Religious beliefs are non-negotiable; the characterization of climate change and related possible policy solutions as this kind of belief divides heroes and villains into two opposing groups with no option of overcoming these established boundaries. It thus diminishes the chances to come to an agreement on the political level. Thus, this narrative stands out from the others as it inadvertently does not allow for a solution to climate change. By contrast, the narrative of responsibility and history or the narrative of positive economic consequences of climate change for example portrays the villain with the option of developing into a hero or – at least – to join the hero in his quest for climate mitigation. Heroes imply in their statements a sense of community by including themselves in the group of perpetrators:

> Countries like the USA or European countries, which are emitting CO^2 beyond the bearable amount for decades; [...] We caused climate change, and those threshold countries are on the best way to catch up with us. But we are responsible for the larger parts of this development that lead to global warming. (I-GER 8: 40)

Thus, the barrier between hero and villain becomes permeably. Both groups are divided by their present attitude towards climate change, on the one side those willing to take action, on the other side those who are ignoring the problem, but a common background and a community allows for a future development where villains turn into heroes or at least join them in their quest.

The narrative of climate change as a partisan issue serves as self-affirmation for the group of climate advocates that they are doing "the right thing" and stand on moral high ground. However, this story of climate change establishes impervious boundaries between two groups.

Since members of both groups need to be included in the quest to lessen the impacts of climate change, this narrative structure achieves the opposite of a unified political elite that tackle the problem of climate change. Here, an even stronger barrier between both groups is created. Thus, the possibilities to make a collective effort in the fight against climate change are narratively diminished. This chapter showed how characters in a narrative relate to one another and how storytellers adjust narratives to subjects they assume are of interest and concern to the audience. It also discussed if and how motifs recur in this study's narrative that were found in other research and thus relates the findings of this study to the existing research.

However, narrative analysis is not only an academic exercise but can also entail suggestions for policy makers and climate advocates, especially in the realm of science communication. Social values, social norms and the perception of social life are a vital factor in adaptation and mitigation policy and in the process of ensuring public support for these policies. From the perspective of the argumentative turn in policy analysis Verweij points out that people's differing and opposing perception of risks and policy challenges need to be understood and used through constructive communication (Verweij et al. 2006: 821). Policy strategies need to consider how risks are communicated in the public discourse, a culturally rich narrative analysis can provide clues as to how public opinion is formed and which inherent structures are at play (Trumbo and Shanahan 2000: 200). Insights from cultural sociology can help here to uncover how narratives interpret data and information about climate change (Osbaldiston 2010: 7).

The analysis of climate change narratives allows for two main policy implications:

1. Contradictory forms of narratives tend to weaken the credibility of those narratives
2. A narrative needs to allow for character development otherwise it will only lead to a deadlock in the process of reaching a consensus.

Ad 1: The study revealed several dominant stories told by climate advocates. The multitude of these stories leads to a contention, not only in the struggle for public attention, but also – and more importantly – in contradictory storylines. Each narrative is built around a different main topic. These topics have varying impacts on the form ($\triangleq$ genre) a narrative takes, leading to inconsistencies between the stories and thus making them vulnerable to attacks. To show how different topics affect the credibility and congruency of narratives, I will turn to the two narratives that are most opposing to each other: the narrative of economy and the narrative of solidarity and humanity.

The latter plays on the value of human equality, insisting that people are obliged to help each other based on sheer humanitarianism. Those who suffer from the impacts of climate change are human as well and thus citizens of the industrialized countries should feel obliged to do something about this suffering. It is the idea of a global, human community that calls for action against climate change. Here, the topic of climate change is moved into a framework of humanitarian sensibility. The stakes here are high and inaction will have devastating consequences.

The former narrative makes use of market rationality, introducing the topic of climate change into the realm of economic arguments. Financial and economic benefits or resp. the avoidance of financial losses serves as motivator for action. However, this leaves a choice for action or inaction: if financial incentives are not going to pay off or if short-term solutions to national economies (for example relying further on high-carbon energy sources) promise to be effective, the argument for getting engaged in climate mitigating activities loses its ground. The topic of climate change and the motivation for taking action is introduced into a mundane setting where money, profit, and economic reason rule. Thus, it is presented as optional and becomes a subject to economic whim.

These two narratives serve as examples for contradictory forms. On the one hand, there is one narrative with high stakes and devastating consequences. On the other hand, there is another narrative where the purpose becomes flexible and needs to meet economic purpose. The final goal is economic prosperity not survival under human conditions. The former is calling for urgent action, the latter settling for a cost-benefit-analysis.

Climate advocates try to address a diverse range of people, which leads them to re-structure the story they are telling, trying to "hit the nerve" of the audience they imagine. Thus, they for example introduce economic reason into the story in order to "get their message across" as one interviewee put it (I-USA 2: 40), they switch from humanitarian beliefs (narrative of solidarity and humanity) to allocation of blame (narrative of responsibility and history). All narratives however deal with the same object. The contradictory forms in which the object is presented now leads to contradictory directions, making the story of climate change action vulnerable to attacks. Policy makers have to consider these contradictions and understand that conflicting stories are not only assigned to antagonistic interest groups, but that they exist within largely unified groups.

Ad 2: The way policy solutions and need for action is presented has to allow for a development of the characters. Only if villains are allowed to turn into heroes, policy agreements can be reached and put into action. Considering for example the narrative of climate change as partisan distinction: here, heroes and villains are set into the opposition of Democrats and Republicans. When former Democratic presidential candidate Al Gore aroused a lot of public attention with the documentary "An inconvenient truth" and earmarked climate change even further as Democratic topic, Republicans had to take an opposing side towards the issue. Considering the multiple legislative and executive levels that are necessary for sustainable climate change policies, a broad political consensus about this problem seems to be important (Engels et al. 2013: 1019; Dunlap, McCright 2008). Constructing strong boundaries between two groups compromises the chances to win people over. The value of the topic is compromised by prejudices of one group towards the advocate group. Thus, the construction of the story of climate change and the layout of policy solutions needs to portray both sides as reasonable to allow the villain to change and join the hero's quest. This prevents a (further) polarization of characters and might ensure public support for policy strategies.

Climate change is a culturally narrated social phenomenon that does not only require the reading of meteorological data but the "reading of those readings" (Smith 2012: 759). The study at hand aimed at revealing this "reading of those readings", i.e. how climate advocates turn climate change into a culturally and socially meaningful story. Outlining an approach to narrative analysis that sufficiently takes into account cultural sociological paradigms, this study presented and analyzed the structure, content, and form of climate change narratives.

Lately, the term "narrative" has become a buzzword, highly popular among policy makers, journalists, motivational speakers and many more. There are constant calls to "change the narrative" or the need for "new narratives". But narratives are more than just a fancy term to spice up a policy speech; the term contains a rich analytical background and great investigative power. But to make fully use of this, we will need to continue working towards an analytically distinct and sharp tool, so we can further our understanding of how societies make sense of the world and how this translates into actions and decision making. Even if this – at first glance – seems like a purely academic problem, at the end it can help advocates and policy makers to develop political and communicative action against global crises, like climate change, but also many more. Not in terms of creating manipulating strategies, but in terms of taking serious the cultural fabric of societies. By understanding how we, as storytelling animals, give meaning to our lives and the world around us. In short, by advancing something, we inherently carry within ourselves: the power of telling and listening to stories, just as we were used to when we were kids.

NOTES

1. Lamont defines boundary work as the process through which people differentiate themselves from others. These patterns of boundary work are not to be understood as individualized characteristics, but as institutionalized cultural repertoires (Lamont 2000): 243.

REFERENCES

Dunlap, R. E., & McCright, A. M. (2008). A Widening Gap: Republican and Democratic Views on Climate Change. *Environment: Science and Policy for Sustainable Development, 50*(5), 26–35.
Engels, A., Hüther, O., Schäfer, M., & Held, H. (2013). Public Climate-change Skepticism, Energy Preferences and Political Participation. *Global Environmental Change, 23*(5), 1018–1027. https://doi.org/10.1016/j.gloenvcha.2013.05.008.
Hart, P., & Nisbet, E. C. (2011). Boomerang Effects in Science Communication: How Motivated Reasoning and Identity Cues Amplify Opinion Polarization About Climate Mitigation Policies. *Communication Research*. https://doi.org/10.1177/0093650211416646.
Jones, M. D. (2010). *Heroes and Villains: Cultural Narratives, Mass Opinions, and Climate Change*. Unpublished Dissertation. University of Oklahoma, Norman.
Lamont, M. (2000). *The Dignity of Working Men. Morality and the Boundaries of Race, Class and Immigration*. Cambridge: Harvard University Press.
Lamont, M., & Fournier, M. (1992). Introduction. In M. Lamont & M. Fournier (Eds.), *Cultivating Differences. Symbolic Boundaries and the Making of Inequality* (pp. 1–20). Chicago: University of Chicago Press.
Liverman, D. M. (2009). Conventions of Climate Change: Constructions of Danger and the Dispossession of the Atmosphere. *Journal of Historical Geography, 35*(2), 279–296.
Maibach, E. W., Nisbet, M. C., Baldwin, P., Akerlof, K., & Diao, G. (2010). Reframing Climate Change as a Public Health Issue: An Exploratory Study of Public Reactions. *BMC Public Health, 10*(1), 299. https://doi.org/10.1186/1471-2458-10-299.

Myers, T. A., Nisbet, M. C., Maibach, E. W., & Leiserowitz, A. A. (2012). A Public Health Frame Arouses Hopeful Emotions About Climate Change. A Letter. *Climatic Change, 113*(3–4), 1105–1112. https://doi.org/10.1007/s10584-012-0513-6.

Ney, S., & Thompson, M. (2000). Cultural Discourses in the Global Climate Change Debate. In E. Jochem, J. A. Sathaye, & D. Bouille (Eds.), *Society, Behaviour, and Climate Change Mitigation* (pp. 65–92). Dordrecht: Springer. Available online at http://link.springer.com/book/10.1007/0-306-48160-X/page/1.

Nisbet, M. C. (2009). Communicating Climate Change. Why Frames Matter for Public Engagement. *Environment: Science and Policy for Sustainable Development, 51*(2), 12–23.

Norgaard, K. M. (2011). *Living in Denial. Climate Change, Emotions, and Everyday Life*. Cambridge, MA: MIT Press.

Osbaldiston, N. (2010). What Role can Cultural Sociology Play in Climate Change Adaptation? Report from the National Climate Change Adaptation Research Facility Conference 2010. In Cultural Sociology Group (Ed.), *Cultural Fields. Newsletter of the TASA Cultural Sociology Thematic Group*, No. 2, pp. 6–7.

Renn, O. (2011). The Social Amplification/Attenuation of Risk Framework. Application to Climate Change. *Wiley Interdisciplinary Reviews: Climate Change, 2*(2), 154–169. https://doi.org/10.1002/wcc.99.

Smith, P. (2012). Narrating Global Warming. In J. C. Alexander, R. N. Jacobs, & P. Smith (Eds.), *The Oxford Handbook of Cultural Sociology* (pp. 745–760). New York: Oxford University Press.

Trumbo, C. W., & Shanahan, J. (2000). Social Research on Climate Change: Where We Have Been, Where We Are, and Where We Might Go. *Public Understanding of Science, 9*(3), 199–204.

Verweij, M., Douglas, M., Ellis, R., Engel, C., Hendriks, F., Lohmann, S., et al. (2006). Clumsy Solutions for a Complex World: The Case of Climate Change. *Public Administration, 84*(4), 817–843. https://doi.org/10.1111/j.1540-8159.2005.09566.x-i1.

Index[1]

A
Alexander, Jeffrey C., 2, 42, 59, 61, 67, 68

C
Climate change communication, 1–4, 7–44, 78, 124–126
Cultural sociology, 38–44, 58, 61, 67, 78, 129
Cultural theory, 3, 27–38, 58, 72–74, 76

D
De Saussure, Ferdinand, 77
Douglas, Mary, 29–31, 72

F
Fear messages, 8, 9, 121n3, 121n4

G
Giddens, Anthony, 10, 11

I
Information deficit model, 20, 22, 41
Issue attention cycle, 15–18

J
Jones, Michael D., 21, 27, 29, 31, 35, 38, 45n13, 58, 62, 64, 67, 72–76, 125

L
Labov, William, 58, 62, 64, 65, 67, 72, 74–76
Linguistics, 14, 58, 59, 64, 67, 74

N
Narratives
  analysis of, 37, 64, 67, 68, 72–74, 76, 77, 124
  characters, 29, 36, 61–63, 70, 73, 77, 83, 86, 96, 115, 123, 125, 129, 131

[1] Note: Page numbers followed by "n" refer to notes

© The Author(s) 2018
A. Arnold, *Climate Change and Storytelling*,
Palgrave Studies in Environmental Sociology and Policy,
https://doi.org/10.1007/978-3-319-69383-5

Narratives (*cont.*)
 conflicting, 2, 34, 130
 defining elements of, 58, 74–76
 and social theory, 74
Narrative theory, 3, 4, 7, 31, 34, 58, 61, 68, 77
Narrative turn, 34, 36, 59, 61
Norgaard, Kaari, 23, 27–29, 35, 125
Norm of balanced reporting, 18, 19, 21, 29

P
Public understanding of science, 7

S
Semiotics, 62, 63, 78n1

Smith, Philip, 19, 27, 29, 35, 38–41, 44, 58, 61, 67–72, 74–76, 78, 124, 125, 131
Structural model, 41, 58, 64–72, 76

V
Value-action gap, 10, 11

W
Waletzky, Joshua, 58, 62, 64, 65, 67, 72, 74–76
White, Hayden, 59–64, 68, 77
Wildavsky, Aaron, 29, 30, 72

CPSIA information can be obtained
at www.ICGtesting.com
Printed in the USA
LVHW071011291119
638297LV00011B/453/P